SpringerBriefs in Mathematics

SpringerBriefs in Mathematics showcases expositions in all areas of mathematics and applied mathematics. Manuscripts presenting new results or a single new result in a classical field, new field, or an emerging topic, applications, or bridges between new results and already published works, are encouraged. The series is intended for mathematicians and applied mathematicians.

More information about this series at http://www.springer.com/series/10030

SBMAC SpringerBriefs

The **SBMAC SpringerBriefs** series publishes relevant contributions in the fields of applied and computational mathematics, mathematics, scientific computing, and related areas. Featuring compact volumes of 50 to 125 pages, the series covers a range of content from professional to academic.

The Sociedade Brasileira de Matemática Aplicada e Computacional (Brazilian Society of Computational and Applied Mathematics, SBMAC) is a professional association focused on computational and industrial applied mathematics. The society is active in furthering the development of mathematics and its applications in scientific, technological, and industrial fields. The SBMAC has helped to develop the applications of mathematics in science, technology, and industry, to encourage the development and implementation of effective methods and mathematical techniques for the benefit of science and technology, and to promote the exchange of ideas and information between the diverse areas of application.

http://www.sbmac.org.br/

Carlile Lavor • Sebastià Xambó-Descamps
Isiah Zaplana

A Geometric Algebra Invitation to Space-Time Physics, Robotics and Molecular Geometry

Springer

Carlile Lavor
Department of Applied Maths
(IMECC-UNICAMP)
University of Campinas
Campinas, SP, Brazil

Sebastià Xambó-Descamps
Departament de Matemàtiques
Universitat Politècnica de Catalunya
Barcelona, Barcelona, Spain

Isiah Zaplana
Institut d'Org. i Control de Sist. Ind.,
Universitat Politècnica de Catalunya
Barcelona, Barcelona, Spain

ISSN 2191-8198 ISSN 2191-8201 (electronic)
SpringerBriefs in Mathematics
ISBN 978-3-319-90664-5 ISBN 978-3-319-90665-2 (eBook)
https://doi.org/10.1007/978-3-319-90665-2

Library of Congress Control Number: 2018940261

Printed on acid-free paper

This Springer imprint is published by the registered company Springer International Publishing AG part of Springer Nature.
The registered company address is: Gewerbestrasse 11, 6330 Cham, Switzerland

Preface

This brief is a pedagogical presentation of key elements of geometric algebra (GA) and a selected sample of research areas in which it is being profitably applied. In spite of its brevity, it is remarkably comprehensive and self-contained. Its contents are designed to be particularly useful to advanced undergraduate students, beginning graduate students, and professionals wishing to understand what are the essential lineaments of GA and how it is effectively used to frame scientific theories and engineering models.

On the applied side, we cover topics in physics, robotics, and molecular geometry. In more detail, the main themes are the geometry and physics of Minkowski's space-time, including Maxwell's electromagnetism and Dirac's equation (Chap. 3); robot forward and inverse kinematics, including an overview of the singularity problem for serial robots (Chap. 4); and protein structure calculations using nuclear magnetic resonance data (Chap. 5).

These three chapters are largely independent, but they require the background material covered in the first two chapters. Chapter 1 contains a detailed study of the geometric algebras of the Euclidean plane and space ($\mathcal{G}_2$ and the Pauli algebra $\mathcal{P} = \mathcal{G}_3$) and of the four-dimensional Minkowski space (the Dirac algebra $\mathcal{D} = \mathcal{G}_{1,3}$). We also illustrate the ways in which $\mathcal{G}_2$ and $\mathcal{G}_3$ encode geometrical facts in their domains. In the case of $\mathcal{D}$, we explain its deep relationship with the Pauli algebra. Chapter 2 is devoted to the conformal geometric algebra (CGA) of the Euclidean space ($\mathcal{C} = \mathcal{G}_{4,1}$), the system required for robotics and molecular geometry.

The relevance of GA today, even for the study of elementary plane and space geometry, deserves a comment. The most straightforward reason that comes to mind is that these geometries are neatly and compactly described, as spelled out in the first chapter, using a formalism that turns out to be equally well adapted to deal with more general fields in mathematics, science, and engineering. In this way, elementary Euclidean geometry provides, particularly for people new to the GA formalism, a tangible path to a realm worthy of being better known by anyone concerned with how best to understand and think about geometry and its applications.

In general, GA provides strong advantages with respect to more common approaches, as for instance that objects and transformations live in the same algebra; that the product in the algebra (geometric product) behaves much like the product of real numbers (except that it is not commutative), which allows easy and familiar manipulations; that the expressions in the algebra are coordinate free; and that it is well adapted to algorithmic and computational treatments.

The reasons why such statements have to be stressed even today have been analyzed by many prominent authors, but most conspicuously and forcefully by D. Hestenes in many of his works, particularly [40] (2nd edition, with a new Preface by the author and a Foreword by A. Lasenby) and [48]. Such enlightened works, and the references provided in them, justify that we do not address here general historical and methodological considerations and that instead we refer the reader to the specific related comments included in the chapters ahead.

The numbering of reusable statements is done in the usual way, like (2.12) for the twelfth equation in Chap. 2 or 2.4.3 for the third numbered statement in Sect. 2.4. The exercises at the end of each chapter are identified in the form E.2.5, which means the fifth exercise in Chap. 2. In the text, these exercises are referred by their identifiers and the page on which they can be found.

A word on the exercises is in order. Usually they are not unlike other items in the text, but with more detail left for the reader to fill in. In such cases, they cover materials that round some aspects of the main text, occasionally including suggestions for streamlined proofs of some standard results or to highlight interesting applications.

Acknowledgements

Carlile Lavor: to the Brazilian Research Agencies CNPq and FAPESP, for their financial support, and to the "Cátedra Ibero-Americana" for June 2017 promoted by the University of Campinas and Polytechnic University of Catalonia.

Sebastià Xambó-Descamps: to the Polytechnic University of Catalonia for its continuous support, and particularly in the last 2 years concerning activities related to geometric algebra; to Jose A. VALLEJO, of the University San Luis Potosí (Mexico), and Felix DELGADO DE LA MATA, Director of the Institute of Mathematics at the University of Valladolid (IMUVA), for inviting me to deliver mini-courses on GA and its applications in 2015 and 2016 and in which several themes developed in Chaps. 1–3 were outlined; and to LA GACETA of the Real Sociedad Matemática Española for allowing us to use the translation into English of the paper [96] as a backbone for Chaps. 1 and 3.

Isiah Zaplana: to the Polytechnic University of Catalonia for its support, especially to the Institute of Industrial and Control Engineering, and to Joan LASENBY, of the University of Cambridge, for introducing me in this fascinating world.

Campinas, SP, Brazil — Carlile Lavor
Barcelona, Spain — Sebastià Xambó-Descamps
Barcelona, Spain — Isiah Zaplana
20 February 2018

Contents

1 Low Dimensional Geometric Algebras 1
1.1 Linear Algebra Background 3
1.2 GA of the Euclidean Plane, $\mathcal{G}_2$ 8
1.3 GA of the Euclidean Space, $\mathcal{G}_3$ 14
1.4 GA of the Minkowski Space, $\mathcal{G}_{1,3}$ 23
1.5 Exercises 30

2 Conformal Geometric Algebra 33
2.1 Ground Notions 33
2.2 Inner Representations 38
2.3 Outer Representations 41
2.4 Transformations 44
2.5 Exercises 49

3 Minkowski's Space-Time: Geometry and Physics 53
3.1 From Physics to Geometry and Back 54
3.2 A GA View of the Lorentz Group 60
3.3 A GA View of Electrodynamics 63
3.4 A GA View of Dirac's Equation 70
3.5 Exercises 72

4 Robot Kinematics 75
4.1 Classical Kinematics 75
4.2 Forward Kinematics 80
4.3 Differential Kinematics 83
4.4 Inverse Kinematics 86
4.5 Identification of Singularities 98
4.6 Exercises 99

5 Molecular Geometry 101
5.1 Distance Geometry 101
5.2 Discretizable Molecular Distance Geometry Problem 102
5.3 Cartesian and Internal Coordinates 104

5.4 Geometric Algebra Approach 106
5.5 Uncertainties from NMR Data 108
5.6 Conformal Geometric Algebra Approach 109
5.7 Exercises 116

References 117

Index 123

Chapter 1
Low Dimensional Geometric Algebras

The concept of *geometric algebra* (GA) arises out of the desire to multiply vectors with the usual rules of multiplying numbers, including the usual rules for taking inverses. From that point of view, the construction of GA is an instance of a powerful mechanism used in mathematics that may be described as *creating virtue out of necessity*. In general, this mechanism comes to the rescue when the need arises to extend a given structure in order to include desirable features that are not present in that structure.

As an aside, and for the purpose of illustrating how the mechanism works in more familiar grounds, let us consider the successive extensions of the notion of number:

$$\mathbb{N} \subset \mathbb{Z} \subset \mathbb{Q} \subset \mathbb{R} \subset \mathbb{C}.$$

In the *natural numbers*, $\mathbb{N} = \{1, 2, 3, \ldots\}$, the difference $x = a - b$ $(a, b \in \mathbb{N})$, which by definition satisfies $a = b + x$, is defined only when $a > b$. The need to be able to subtract any two numbers leads to the introduction of 0 and the *negative numbers* $-a$ (for all $a \in \mathbb{N}$). The extension $\mathbb{Z} = \{\ldots, -3, -2, -1, 0, 1, 2, 3, \ldots\}$ of $\mathbb{N}$ is the set of *integers*. The order, addition, and subtraction can be extended in a natural way from $\mathbb{N}$ to $\mathbb{Z}$ and after that the equation $a = b + x$ has a unique solution x for any $a, b \in \mathbb{Z}$. In other words, the *difference* $x = a - b$ of any two integers is always well defined. For $a, b \in \mathbb{N}$, for example, $x = a - b$ if $a > b$, 0 if $a = b$, $-(b - a)$ if $a < b$. The bottom line is that $\mathbb{Z}$ implements the order, addition, and subtraction of *formal differences* $a - b$ of natural numbers a and b, with the constraints that $a - b = a' - b'$ (respectively $a - b < a' - b'$) if and only if $a + b' = a' + b$ (respectively $a + b' < a' + b$) in $\mathbb{N}$.

Now the division $x = a/b$ $(b \neq 0)$ is possible in $\mathbb{Z}$ precisely when b is a divisor of a. In fact, to say that the equation $a = bx$ can be solved for $x \in \mathbb{Z}$ just says that a is a multiple of b. The wish to overcome this limitation of the integers leads to the

C. Lavor et al., *A Geometric Algebra Invitation to Space-Time Physics, Robotics and Molecular Geometry*, SpringerBriefs in Mathematics,
https://doi.org/10.1007/978-3-319-90665-2_1

introduction of *fractions* or *rational numbers*, $\mathbb{Q} = \{a/b : a, b \in \mathbb{Z},\ b \neq 0\}$. The bottom line here is that $\mathbb{Q}$ implements the order, addition, and subtraction of *formal quotients* a/b of integers a and b ($b \neq 0$), with the constraints that $a/b = a'/b'$ (respectively $a/b < a'/b'$) if and only if $ab' = a'b$ (respectively $ab' < a'b$) in $\mathbb{Z}$. In particular, the equation $a = bx$ ($a, b \in \mathbb{Z}, b \neq 0$) can be solved uniquely for x in $\mathbb{Q}$: $x = a/b$.

The real numbers $\mathbb{R}$ can be introduced as the natural extension of $\mathbb{Q}$ that makes possible to take the *least upper bound* of *upper bounded sets*, and $\mathbb{C}$ is the natural extension of $\mathbb{R}$ in which -1 *has a square root*: $i = \sqrt{-1}$. Operationally, the *number* i is manipulated so that $i^2 = -1$.

At this point it is worthwhile to remark that the mechanism is more fertile than what it might appear at a first sight. As a rule, the new structures obtained to overcome some limitations of more primitive ones have a richness that goes far beyond the original specifications, both by having interesting unexpected features and for its capacity to suggest other potentially useful structures through analogy and generalization. We will see this at work throughout this brief, notably in the case of geometric algebra, and also in a number of scattered comments.

Here it should be sufficient to recall a couple of examples. Given any positive number $a \in \mathbb{R}$, it turns out that it has a unique positive real nth root $r = \sqrt[n]{a}$ for any $n \in \mathbb{N}$ (this means that $r^n = a$), a fact that is not true in $\mathbb{Q}$, as reminded by the old Pythagoric story telling us that $\sqrt{2}$ cannot be a rational number (see E.1.1, p. 30).

In the same vein, when we accepted that there is i such that $i^2 = -1$, and thus extending $\mathbb{R}$ to $\mathbb{C}$, how could we suspect that for any non-zero $z \in \mathbb{C}$, and $n \in \mathbb{N}$, there are exactly n numbers $\xi \in \mathbb{C}$ such that $\xi^n = z$ (nth roots of z)? For example, (see E.1.2, p. 31), the nth roots of 1 are $e^{2\pi ik/n} = \cos(2\pi k/n) + i\sin(2\pi k/n)$ ($0 \leqslant k < n$).

In the case of GA, among the unexpected properties beyond its specification (which is the wish to multiply vectors as if they were numbers), we will find that it is *capable of representing in a coordinate-free way both geometrical concepts and geometric operations on them*. Moreover, these two roles are naturally related in a way that will be made precise in due time and which we call *geometric covariance*.

The aim of this chapter is to introduce and study some of the concrete geometric algebras that will be used in the remaining chapters. These include the geometric algebras $\mathcal{G}_2$ and $\mathcal{G}_3$ of the Euclidean plane E_2 and the Euclidean space E_3 (Sections 2 and 3, respectively) and the geometric algebra $\mathcal{G}_{1,3}$ of the Minkowski space $E_{1,3}$. To pave the way to later chapters, in the Euclidean cases we also provide details about how the $\mathcal{G}_2$ and $\mathcal{G}_3$ encode geometric notions and geometric transformations.

Convention. If f is a map and $\boldsymbol{x}$ an object (say a linear map and a vector), we allow ourselves to (optionally) write $f\boldsymbol{x}$, instead of $f(\boldsymbol{x})$, to denote the image of $\boldsymbol{x}$ by f. This device, which is a common practice in functional programming languages, is useful to increase the readability of expressions in contexts where no confusion can arise about the nature of f and $\boldsymbol{x}$.

1.1 Linear Algebra Background

We assume that the reader is familiar with some basic notions of linear algebra. For reference convenience, here is a summary of what we need in the sequel.

By a *vector space* we mean a real vector space (also called an $\mathbb{R}$-vector space). The elements of $\mathbb{R}$ are called *scalars* and will be denoted by Greek letters: $\alpha, \beta, \ldots$.

In each concrete case of geometric algebra, the starting point is a vector space E of finite dimension n (with $n \leqslant 4$ in this chapter). Its elements are denoted by boldface italic characters ($\boldsymbol{e}, \boldsymbol{u}, \boldsymbol{v}, \boldsymbol{x}, \boldsymbol{y}, \ldots$).

The vector subspace of E spanned by vectors $\boldsymbol{x}_1, \ldots, \boldsymbol{x}_k$ (that is, the set of all linear combinations $\lambda_1\boldsymbol{x}_1 + \cdots + \lambda_k\boldsymbol{x}_k$, $\lambda_1, \ldots, \lambda_k \in \mathbb{R}$) will be denoted by $\langle \boldsymbol{x}_1, \ldots, \boldsymbol{x}_k \rangle$.

The symbol $\mathbf{e}$ will stand for a basis $\boldsymbol{e}_1, \ldots, \boldsymbol{e}_n$ of E. In principle it is arbitrary, but often it will be assumed to have specific properties that in each case will be declared explicitly.

If E' is another vector space, a map $f : E \to E'$ is said to be *linear* if $f(\lambda\boldsymbol{x}) = \lambda f\boldsymbol{x}$ and $f(\boldsymbol{x} + \boldsymbol{y}) = f\boldsymbol{x} + f\boldsymbol{y}$ for all $\boldsymbol{x}, \boldsymbol{y} \in E$ and $\lambda \in \mathbb{R}$.

1.1.1 (Construction of linear maps) The main device to construct linear maps is the following observation: *If we are given any vectors $\boldsymbol{e}'_1, \ldots, \boldsymbol{e}'_n \in E'$, then there is a* ***unique*** *linear map $f : E \to E'$ such that $f\boldsymbol{e}_j = \boldsymbol{e}'_j$ for $j = 1, \ldots, n$.* □

Metrics

The vector space E on which geometric algebra is grounded is supposed to be equipped with a *metric*. By this we understand a *non-degenerate* (or *regular*) symmetric bilinear form $q : E \times E \to \mathbb{R}$. Recall that the non-degenerate condition means that for any given vector $\boldsymbol{x} \neq 0$ we can find a vector $\boldsymbol{y}$ such that $q(\boldsymbol{x}, \boldsymbol{y}) \neq 0$ or, equivalently, $q(\boldsymbol{x}, \boldsymbol{y}) = 0$ for all $\boldsymbol{y}$ implies that $\boldsymbol{x} = 0$. Instead of $q(\boldsymbol{x}, \boldsymbol{x})$, which is the *quadratic form* associated with q, we will simply write $q(\boldsymbol{x})$. Note that $q(\lambda\boldsymbol{x}) = \lambda^2 q(\boldsymbol{x})$.

1.1.2 (Polarization identity) *The quadratic form determines the metric:*

$$2q(\boldsymbol{x}, \boldsymbol{y}) = q(\boldsymbol{x} + \boldsymbol{y}) - q(\boldsymbol{x}) - q(\boldsymbol{y}) \quad \textit{for all} \quad \boldsymbol{x}, \boldsymbol{y} \in E.$$

Proof Use the bilinear property and the symmetry of q:

$$\begin{aligned} q(\boldsymbol{x} + \boldsymbol{y}) &= q(\boldsymbol{x} + \boldsymbol{y}, \boldsymbol{x} + \boldsymbol{y}) = q(\boldsymbol{x}, \boldsymbol{x}) + q(\boldsymbol{x}, \boldsymbol{y}) + q(\boldsymbol{y}, \boldsymbol{x}) + q(\boldsymbol{y}, \boldsymbol{y}) \\ &= q(\boldsymbol{x}) + q(\boldsymbol{y}) + 2q(\boldsymbol{x}, \boldsymbol{y}). \end{aligned}$$ □

Two vectors $\boldsymbol{x}, \boldsymbol{y} \in E$ are said to be *orthogonal* precisely when $q(\boldsymbol{x}, \boldsymbol{y}) = 0$. The basis $\mathbf{e}$ is orthogonal if $q(\boldsymbol{e}_j, \boldsymbol{e}_k) = 0$ for all $j \neq k$. As is well known, and easy to proof, orthogonal basis exist for any q (E.1.3, p. 31).

A *q-isometry* of E (or just *isometry* if q is understood from the context) is a linear map $f : E \to E$ such that $q(f\boldsymbol{v}, f\boldsymbol{v}') = q(\boldsymbol{v}, \boldsymbol{v}')$ for all $\boldsymbol{v}, \boldsymbol{v}' \in E$. Using the polarization identity 1.1.2, we see that *f is an isometry if and only if $q(f\boldsymbol{v}) = q(\boldsymbol{v})$ for all vectors $\boldsymbol{v}$*. With the operation of composition, the set of q-isometries forms a group, with the identity map Id as its neutral element. It is called the *orthogonal group* of q and is denoted by O_q.

Note that *an isometry maps orthogonal vectors to orthogonal vectors.*

Euclidean Spaces

Many authors take $\mathbb{R}^n$ as a model for the n-dimensional Euclidean vector space, but we prefer to denote it E_n (or E if the dimension is clear from the context) to stress that no basis is assigned a preferred role. So E_n is a real vector space of dimension n endowed with an *Euclidean metric* q, which means that $q(\boldsymbol{v}) = q(\boldsymbol{v}, \boldsymbol{v}) > 0$ for any non-zero vector $\boldsymbol{v}$ (note that $q(0) = 0$ follows from the bilinearity of q). We also say that q is *positive definite*.

The *length* (or *norm*) of a vector $\boldsymbol{v}$ is denoted by $|\boldsymbol{v}|$ and is defined by the formula

$$|\boldsymbol{v}| = \sqrt{q(\boldsymbol{v})}. \tag{1.1}$$

Thus $|\boldsymbol{v}| > 0$ for $\boldsymbol{v} \neq 0$ and $q(\boldsymbol{v}) = |\boldsymbol{v}|^2$ for any $\boldsymbol{v}$.

A vector $\boldsymbol{u}$ such that $q(\boldsymbol{u}) = 1$ is said to be a *unit vector*. For any non-zero vector $\boldsymbol{v}$, $\pm\boldsymbol{v}/|\boldsymbol{v}|$ are the only unit vectors of the form $\lambda\boldsymbol{v}$ ($\lambda \in \mathbb{R}$), and $\boldsymbol{v}/|\boldsymbol{v}|$ is said to be the *normalization* of $\boldsymbol{v}$.

Notice that if we normalize the vectors of an orthogonal basis we get an orthogonal basis of unit vectors. Such bases are said to be *orthonormal*.

The *angle* $\alpha = \alpha(\boldsymbol{v}, \boldsymbol{v}') \in [0, \pi]$ between two non-zero vectors $\boldsymbol{v}$ and $\boldsymbol{v}'$ is the real number α defined by the relation

$$\cos\alpha = q(\boldsymbol{v}, \boldsymbol{v}')/|\boldsymbol{v}||\boldsymbol{v}'|. \tag{1.2}$$

By the Cauchy-Schwarz inequality (see E.1.4, p. 31), this is well defined. Moreover, $\alpha \in (0, \pi)$ when $\boldsymbol{v}$ and $\boldsymbol{v}'$ are linearly independent and $\alpha = 0$ ($\alpha = \pi$) when $\boldsymbol{v}' = \lambda\boldsymbol{v}$ with $\lambda > 0$ ($\lambda < 0$) . Note also that $\alpha = \pi/2$ if and only if $q(\boldsymbol{v}, \boldsymbol{v}') = 0$, that is, if and only if $\boldsymbol{v}$ and $\boldsymbol{v}'$ are *orthogonal* (in the Euclidean case, the term *perpendicular* may be used instead).

The isometry group of E_n (*orthogonal group*) will be denoted O_n.

Algebras

By an *algebra* we understand a *non-zero* vector space $\mathcal{A}$ endowed with a bilinear product $\mathcal{A}\times\mathcal{A}\to\mathcal{A}$, $(x,y)\mapsto x*y$. Unless declared explicitly otherwise, we also assume that the product is *associative*, $(x*y)*z=x*(y*z)$, and *unital* (that is, there is $1_{\mathcal{A}}\in\mathcal{A}$ such that $1_{\mathcal{A}}\neq 0_{\mathcal{A}}$ and $1_{\mathcal{A}}*x=x*1_{\mathcal{A}}=x$ for all $x\in\mathcal{A}$).

1.1.3 (Example: The matrix algebra) For any positive integer n, the vector space $\mathbb{R}(n)$ of $n\times n$ real matrices is an algebra with the usual matrix product. Its unit is the matrix I_n that has 1 in the main diagonal and 0 elsewhere *(identity matrix).*

Later we will use the following observation about $\mathbb{R}(2)$. Let

$$\mathfrak{e}_1=\begin{pmatrix}1&0\\0&-1\end{pmatrix},\ \mathfrak{e}_2=\begin{pmatrix}0&1\\1&0\end{pmatrix}.$$

Then it is easily checked that $\{I_2,\mathfrak{e}_1,\mathfrak{e}_2,\mathfrak{e}_1\mathfrak{e}_2\}$ is a basis of $\mathbb{R}(2)$ and that the following relations hold: $\mathfrak{e}_1^2=\mathfrak{e}_2^2=I_2$, $\mathfrak{e}_1\mathfrak{e}_2+\mathfrak{e}_2\mathfrak{e}_1=0$.

The map $\mathbb{R}\to\mathcal{A}$, $\lambda\mapsto\lambda 1_{\mathcal{A}}$, allows us to regard $\mathbb{R}$ as embedded in $\mathcal{A}$, and so we will not distinguish between $\lambda\in\mathbb{R}$ and $\lambda 1_{\mathcal{A}}\in\mathcal{A}$.

Exterior Powers and Exterior Algebra

The exterior powers and the exterior algebra of E, $\wedge^k E$ and $\wedge E$, were discovered by H. Grassmann [37–39]. They do not depend on the metric q, but we have postponed its recall because they are a little more abstract, and this should not hide that they have a clear geometric meaning and are quite manageable in practice. In any case, we will use the exterior algebra to determine the (graded) linear structure of geometric algebras and other related concepts.

Let E^k ($k\in\mathbb{N}$) be the kth *Cartesian power* of E. It is the vector space whose elements are k-tuples of vectors $(\boldsymbol{x}_1,\ldots,\boldsymbol{x}_k)$. The *exterior power* $\wedge^k E$ ($1\leqslant k\leqslant n$) is a vector space endowed with a *skew-symmetric multilinear* map

$$\wedge: E^k\to\wedge^k E,\quad (\boldsymbol{x}_1,\ldots,\boldsymbol{x}_k)\mapsto\boldsymbol{x}_1\wedge\cdots\wedge\boldsymbol{x}_k.$$

Recall that a map is multilinear if it is linear in each of its variables, for an arbitrary value of the remaining variables, and that it is skew-symmetric if it changes sign when any two consecutive variables are swapped. The elements of $\wedge^k E$ are called *k-vectors* and a *k-blade* is a non-zero k-vector of the form $\boldsymbol{x}_1\wedge\cdots\wedge\boldsymbol{x}_k$ (k-blades are also called *decomposable* k-vectors). It is to be thought as the *oriented k-volume* determined by the vectors $\boldsymbol{x}_1,\ldots,\boldsymbol{x}_k$ (oriented *area* and *volume* for $k=2$ and $k=3$). Algebraically, the skew-symmetric condition is reflected by the fact that

$\boldsymbol{x}_1 \wedge \cdots \wedge \boldsymbol{x}_k$ vanishes if and only if $\boldsymbol{x}_1, \ldots, \boldsymbol{x}_k$ are linearly dependent. The restriction $k \leqslant n$ arises from the fact that there are no non-zero k-volumes for $k > n$, a point that can also be expressed by declaring that $\wedge^k E = \{0\}$ for $k > n$. For $k = 1$, $\wedge^1 E = E$ and $\wedge : E \to \wedge^1 E$ is the identity, while $\wedge^0 E = \mathbb{R}$ by convention.

1.1.4 (Universal property) The fundamental property of $\wedge^k E$ is that for any skew-symmetric k-multilinear map $f : E^k \to F$ (where F is any vector space) *there exists a unique* ***linear*** *map* $f^\wedge : \wedge^k E \to F$ such that

$$f(\boldsymbol{x}_1, \ldots, \boldsymbol{x}_k) = f^\wedge(\boldsymbol{x}_1 \wedge \cdots \wedge \boldsymbol{x}_k).$$

The *exterior algebra* (or *Grassmann algebra*) associated with E, $(\wedge E, \boldsymbol{\wedge})$, is the direct sum of the *exterior powers* $\wedge^k E$ of E $(0 \leqslant k \leqslant n)$,

$$\wedge E = \bigoplus_{k=0}^n \wedge^k E = \mathbb{R} \oplus E \oplus \wedge^2 E \oplus \cdots \oplus \wedge^n E,$$

endowed with the exterior product $\boldsymbol{\wedge}$ whose basic computational rule is

$$(\boldsymbol{x}_1 \wedge \cdots \wedge \boldsymbol{x}_k) \boldsymbol{\wedge} (\boldsymbol{y}_1 \wedge \cdots \wedge \boldsymbol{y}_{k'}) = \boldsymbol{x}_1 \wedge \cdots \wedge \boldsymbol{x}_k \wedge \boldsymbol{y}_1 \wedge \cdots \wedge \boldsymbol{y}_{k'}.$$

So it is a *graded algebra*, as $x \boldsymbol{\wedge} x' \in \wedge^{k+k'} E$ when $x \in \wedge^k E$ and $x' \in \wedge^{k'} E$. The exterior product is *skew-commutative* (or *supercommutative*):

$$x \wedge x' = (-1)^{kk'} x' \wedge x,$$

if $x \in \wedge^k E$, $x' \in \wedge^{k'} E$. On account of the associativity of the exterior product, the distinction between $\wedge$ and $\boldsymbol{\wedge}$ is unnecessary and it will not be done in what follows.

The elements of $\wedge E$ are called *multivectors*. Given a multivector $x \in \wedge E$, there is a unique decomposition $x = x_0 + x_1 + \cdots + x_n$ with $x_k \in \wedge^k E$ $(k = 0, 1, \ldots, n)$ and we say that x_k is the *grade k component* of x.

1.1.5 (The parity involution) *If* $x = \sum x_k$ $(x_k \in \wedge^k E)$ *is a multivector, we define* $\hat{x} = \sum (-1)^k x_k$. *This gives a linear map* $\wedge E \to \wedge E$, $x \mapsto \hat{x}$, *that is an involution* (which means that $\hat{\hat{x}} = x$ for all $x \in \wedge E$). *It is called the* parity involution *of* $\wedge E$. *Since it satisfies* $\widehat{x \wedge y} = \hat{x} \wedge \hat{y}$ *for all* $x, y \in \wedge E$, *we say that it is an algebra* automorphism *of* $\wedge E$.

Proof It is a direct consequence of the definitions and the fact that the exterior product is graded. □

1.1.6 (The reverse involution) *If* $x = \sum x_k$ $(x_k \in \wedge^k E)$ *is a multivector, we define* $\tilde{x} = \sum (-1)^{k/\!/2} x_k$ $(k/\!/2 = \lfloor q/2 \rfloor)$. *The linear map* $\wedge E \to \wedge E$, $x \mapsto \tilde{x}$, *is an involution that is called the* reverse involution *of* $\wedge E$. *Since it satisfies* $\widetilde{x \wedge y} = \tilde{y} \wedge \tilde{x}$ *for all* $x, y \in \wedge E$, *we say that it is an algebra* antiautomorphism *of* $\wedge E$.

Proof The key point is that if $x = \boldsymbol{x}_1 \wedge \cdots \wedge \boldsymbol{x}_k$ is a k-blade, then $\tilde{x} = \boldsymbol{x}_k \wedge \cdots \wedge \boldsymbol{x}_1$. Indeed, since the exterior product is skew-symmetric, $\boldsymbol{x}_k \wedge \cdots \wedge \boldsymbol{x}_1 = (-1)^{\binom{k}{2}} x$ and $(-1)^{\binom{k}{2}} = (-1)^{k/\!/2}$ because $\binom{k}{2}$ has the same parity as $k/\!/2$. With this, the proof is easily completed. □

1.1.7 (Grassmann basis) *For any basis* $\boldsymbol{e}_1, \ldots, \boldsymbol{e}_n$ *of* E*, the* $\binom{n}{k}$ *products*

$$\boldsymbol{e}_{\hat{J}} = \boldsymbol{e}_{j_1} \wedge \cdots \wedge \boldsymbol{e}_{j_k} \quad (J = \{j_1, \ldots, j_k\},\ 1 \leqslant j_1 < \ldots < j_k \leqslant n)$$

form a basis of $\wedge^k E$*. In particular,* $\dim \wedge^k E = \binom{n}{k}$ *and* $\dim \wedge E = 2^n$*. Moreover,*

$$\boldsymbol{e}_{\hat{I}} \wedge \boldsymbol{e}_{\hat{J}} = \begin{cases} 0 & \text{if } I \cap J \neq \emptyset \\ (-1)^{t(I,J)} \boldsymbol{e}_{I+J} & \text{otherwise} \end{cases}$$

where $I + J$ *is the result of reordering the concatenated sequence* I, J *in increasing order and* $t(I, J)$ *is the number order inversions in* I, J. □

Proof From the universal property it can be seen that $\wedge^k E$ is spanned by the k-blades. Owing to the skew-symmetric character of the exterior product, $\wedge^k E$ is also spanned by the $\boldsymbol{e}_{\hat{J}}$ in the statement with $|J| = k$. So it will be enough to show that these $\boldsymbol{e}_{\hat{J}}$ are linearly independent. In turn, this can be easily established if we show that for each J there is a linear map $\omega_J : \wedge^k E \to \mathbb{R}$ such that $\omega_J(\boldsymbol{e}_{\hat{K}}) = \delta_{J,K}$ (so 0 for $K \neq J$ and 1 for $K = J$). Now we can produce ω_J by letting $\omega_j : E \to \mathbb{R}$ be the (unique) linear map such that $\omega_j(\boldsymbol{e}_i) = \delta_{i,j}$ and by considering the map $f_J : E^k \to \mathbb{R}$ such that $(\boldsymbol{x}_1, \ldots, \boldsymbol{x}_k) \mapsto \det(\omega_{j_l}(\boldsymbol{x}_i))$, where $1 \leqslant i, l \leqslant k$. The map f_J is skew-symmetric and multilinear, by the properties of the determinant, and it is enough to take as $\omega_J : \wedge^k E \to \mathbb{R}$ the unique linear map such that $\omega_J(\boldsymbol{x}_1 \wedge \cdots \wedge \boldsymbol{x}_k) = f_J(\boldsymbol{x}_1, \ldots, \boldsymbol{x}_k)$. □

The Projective Space P*E*

The projective space $\mathbf{P}E$ of the vector space E is *the set of* 1*-dimensional linear subspaces of* E. To distinguish between the subspace $\langle \boldsymbol{e} \rangle$ ($\boldsymbol{e} \in E - \{0\}$) as a subset of E and as a point of $\mathbf{P}E$, the latter will be denoted $|\boldsymbol{e}\rangle$ (corresponding to the notation $[\boldsymbol{e}]$ in the projective geometry texts). Thus we have $|\boldsymbol{e}\rangle = |\boldsymbol{e}'\rangle$ if and only if $\boldsymbol{e}' = \lambda \boldsymbol{e}$ for some $\lambda \in \mathbb{R}$ (necessarily non-zero because $\boldsymbol{e}, \boldsymbol{e}' \in E - \{0\}$), a relation that henceforth will be written $\boldsymbol{e}' \sim \boldsymbol{e}$.

1.1.8 (Linear subspaces and blades) Let $X = \boldsymbol{x}_1 \wedge \cdots \wedge \boldsymbol{x}_k$ be a k-blade. Then the relation

$$\boldsymbol{x} \in \langle \boldsymbol{x}_1, \ldots, \boldsymbol{x}_k \rangle \Leftrightarrow \boldsymbol{x} \wedge X = 0 \tag{1.3}$$

shows that X determines the subspace $\langle \boldsymbol{x}_1, \ldots, \boldsymbol{x}_k \rangle$. Moreover, if $X' = \boldsymbol{x}'_1 \wedge \cdots \wedge \boldsymbol{x}'_k$ is another k-blade, then $\langle \boldsymbol{x}_1, \ldots, \boldsymbol{x}_k \rangle = \langle \boldsymbol{x}'_1, \ldots, \boldsymbol{x}'_k \rangle$ if and only if $X' \sim X$. The if part follows immediately from (1.3). The only if part is a consequence of the fact that if $F = \langle \boldsymbol{x}_1, \ldots, \boldsymbol{x}_k \rangle = \langle \boldsymbol{x}'_1, \cdots, \boldsymbol{x}'_k \rangle$, then $X, X' \in \wedge^k F$ and hence $X' \sim X$ because $\wedge^k F$ is 1-dimensional. Thus we see that the equality $|X\rangle = |X'\rangle$ in $\mathbf{P}(\wedge^k E)$ is equivalent to the equality of the corresponding subspaces. Because of this, we will denote by $|X\rangle$ the linear subspace $\langle \boldsymbol{x}_1, \ldots, \boldsymbol{x}_k \rangle$ determined by X. □

1.2 GA of the Euclidean Plane, $\mathcal{G}_2$

Let us start our journey with an Euclidean plane E_2. In the first part of this section we will explain some basic notions about the geometry of this plane and in the second part we will introduce $\mathcal{G}_2$ and study how it relates to the geometry.

1.2.1 (Example: The symmetry $s_{\boldsymbol{v}}$) *Let $\boldsymbol{v}$ be a non-zero vector. Then there is a unique $f \in \mathrm{O}_2$, $f \neq \mathrm{Id}$, such that $f(\boldsymbol{v}) = \boldsymbol{v}$. This isometry is called the* symmetry *with respect to $\boldsymbol{v}$, is denoted by $s_{\boldsymbol{v}}$, and it has the property that $s_{\boldsymbol{v}}(\boldsymbol{v}') = -\boldsymbol{v}'$ when $\boldsymbol{v}'$ is perpendicular to $\boldsymbol{v}$* (see Fig. 1.1a).

Proof Indeed, the linear subspace $\boldsymbol{v}^\perp = \{\boldsymbol{v}' \in E_2 \mid q(\boldsymbol{v}, \boldsymbol{v}') = 0\}$ has dimension 1 (it is the kernel of the non-zero linear map $E_2 \to \mathbb{R}$ such that $\boldsymbol{v}' \mapsto q(\boldsymbol{v}, \boldsymbol{v}')$) and f maps $\boldsymbol{v}^\perp$ to itself. If it were $f(\boldsymbol{v}') = \boldsymbol{v}'$ for some non-zero $\boldsymbol{v}' \in \boldsymbol{v}^\perp$, then f would be the identity. Therefore $f(\boldsymbol{v}') = \lambda \boldsymbol{v}'$ for some scalar $\lambda \neq 1$. But $q(\boldsymbol{v}') = q(f\boldsymbol{v}') = \lambda^2 q(\boldsymbol{v}')$ implies that $\lambda^2 = 1$ and hence $\lambda = -1$. □

By *symmetry* we understand a symmetry with respect to a non-zero vector.

1.2.2 (Cartan-Dieudonné theorem for E_2) *Let $f \in \mathrm{O}_2$ be an isometry. Then f is a symmetry or a composition of two symmetries.*

Proof If $f = \mathrm{Id}$, then $f = s^2$ for any symmetry s. So we may assume that $f \neq \mathrm{Id}$. In that case, $f = s_{\boldsymbol{v}}$ if there is a non-zero vector $\boldsymbol{v}$ such that $f(\boldsymbol{v}) = \boldsymbol{v}$. So we may assume that f leaves no vector fixed. If $f(\boldsymbol{v}) = -\boldsymbol{v}$ for all $\boldsymbol{v}$, then $f = -\mathrm{Id} = s_{\boldsymbol{v}} s_{\boldsymbol{v}'}$ for any two non-zero perpendicular vectors $\boldsymbol{v}$ and $\boldsymbol{v}'$. Thus we may assume that there is a vector $\boldsymbol{v}$ such that $f(\boldsymbol{v}) \neq -\boldsymbol{v}$, or $\boldsymbol{v}' = f(\boldsymbol{v}) + \boldsymbol{v} \neq 0$. Then

$$s_{\boldsymbol{v}'}(f(\boldsymbol{v}) + \boldsymbol{v}) = f(\boldsymbol{v}) + \boldsymbol{v} \text{ and } s_{\boldsymbol{v}'}(f(\boldsymbol{v}) - \boldsymbol{v}) = -f(\boldsymbol{v}) + \boldsymbol{v},$$

as $f(\boldsymbol{v}) - \boldsymbol{v}$ is perpendicular to $f(\boldsymbol{v}) + \boldsymbol{v}$ (see Fig. 1.1b). Consequently $s_{\boldsymbol{v}'}(f(\boldsymbol{v})) = \boldsymbol{v}$ and so $s_{\boldsymbol{v}'} f = s_{\boldsymbol{v}}$. Therefore $f = s_{\boldsymbol{v}'} s_{\boldsymbol{v}}$. □

An isometry that is the composition of two symmetries is called a *rotation*. So we have $\mathrm{O}_2 = \mathrm{O}_2^- \sqcup \mathrm{O}_2^+$ (disjoint union), where O_2^- denotes the set of symmetries and O_2^+ the set of rotations. We will see that O_2^+ is a subgroup of O_2 called *special orthogonal group* and which is also denoted by SO_2.

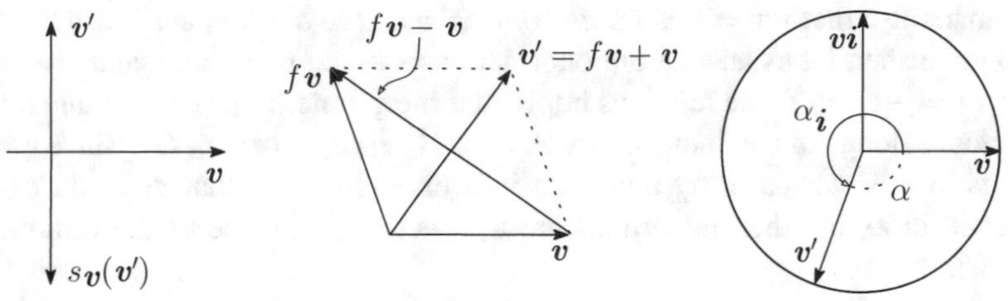

Fig. 1.1 (**a**) Symmetry $s_{\boldsymbol{v}}$. (**b**) An isometry without fixed non-zero vectors is the composition of two symmetries. (**c**) Oriented angle, case $\alpha_i = 2\pi - \alpha$ (see 1.2.6)

Geometric Algebra

We want to enlarge E_2 to a system $\mathcal{G}_2$ in which vectors can be multiplied, and non-zero vectors inverted, with the usual rules. Our strategy will be to show, under some natural assumptions, that $\mathcal{G}_2$ is unique, and then establish its existence by displaying a system that satisfies all the requirements.

Let us start with some general remarks. By xy we denote the product of $x, y \in \mathcal{G}_2$ (simple juxtaposition of the factors) and we say that it is the *geometric product* of x and y. This notion was introduced by Clifford with this name [19], but some authors call it the *Clifford product* (see, for instance, [72] or [92]).

Technically, the structure of $\mathcal{G}_2$ with the geometric product is supposed to be an *associative* and *unital* $\mathbb{R}$-algebra. Recall that this means that $\mathcal{G}_2$ is a real vector space and that the geometric product is bilinear, associative, with unit $1 \in \mathcal{G}_2$. We have $\mathbb{R} \subset \mathcal{G}_2$ and its elements are the *scalars* of $\mathcal{G}_2$. We also have $E_2 \subset \mathcal{G}_2$, and its elements are the *vectors* of $\mathcal{G}_2$. We assume that $\mathbb{R} \cap E_2 = \{0\}$.

Let $\boldsymbol{v} \in E_2$. If $\boldsymbol{v}$ is to have an inverse $\boldsymbol{v}'$ with respect to the geometric product, it is natural to assume that $\boldsymbol{v}' \in \langle \boldsymbol{v} \rangle = \{\lambda \boldsymbol{v} \mid \lambda \in \mathbb{R}\}$, as $\langle \boldsymbol{v} \rangle$ is the only subset of E_2 naturally associated with $\boldsymbol{v}$ using the linear structure. But $1 = \boldsymbol{v}'\boldsymbol{v} = (\lambda \boldsymbol{v})\boldsymbol{v} = \lambda \boldsymbol{v}^2$, which implies that $\boldsymbol{v}^2$ *must be a non-zero scalar.*

If now $\boldsymbol{v}, \boldsymbol{w} \in E_2$, then $(\boldsymbol{v}+\boldsymbol{w})^2 = \boldsymbol{v}^2+\boldsymbol{w}^2+\boldsymbol{v}\boldsymbol{w}+\boldsymbol{w}\boldsymbol{v}$ implies that $\boldsymbol{v}\boldsymbol{w}+\boldsymbol{w}\boldsymbol{v} \in \mathbb{R}$. Thus the expression $\boldsymbol{v}\boldsymbol{w} + \boldsymbol{w}\boldsymbol{v}$ defines a bilinear symmetric product in E_2 and the main insight of Clifford was to postulate the relation

$$\boldsymbol{v}\boldsymbol{w} + \boldsymbol{w}\boldsymbol{v} = q(\boldsymbol{v}, \boldsymbol{w}) + q(\boldsymbol{w}, \boldsymbol{v}) = 2q(\boldsymbol{v}, \boldsymbol{w}) \tag{1.4}$$

(which we will call *Clifford relation*). Setting $\boldsymbol{w} = \boldsymbol{v}$, we get *Clifford's reduction rule*:

$$\boldsymbol{v}^2 = q(\boldsymbol{v}). \tag{1.5}$$

This formula tells us that any non-zero vector $\boldsymbol{v}$ has an inverse $\boldsymbol{v}^{-1}$ with respect to the geometric product and that $\boldsymbol{v}^{-1} = \boldsymbol{v}/q(\boldsymbol{v})$. In particular we have $\boldsymbol{u}^{-1} = \boldsymbol{u}$ if $\boldsymbol{u}$ is a unit vector.

Remark also that $\boldsymbol{v}\boldsymbol{w} = -\boldsymbol{w}\boldsymbol{v}$ *if and only if* $\boldsymbol{v}$ *and* $\boldsymbol{w}$ *are perpendicular.*

To go further, let us take an orthonormal basis $\boldsymbol{e}_1, \boldsymbol{e}_2 \in E_2$, so that $\boldsymbol{e}_1^2 = \boldsymbol{e}_2^2 = 1$ and $\boldsymbol{e}_2\boldsymbol{e}_1 = -\boldsymbol{e}_1\boldsymbol{e}_2$. These relations imply that the geometric product of any number of vectors belongs to the linear subspace $L = \langle 1, \boldsymbol{e}_1, \boldsymbol{e}_2, \boldsymbol{e}_1\boldsymbol{e}_2 \rangle \subseteq \mathcal{G}_2$. Since a vector belongs to this subspace, by induction it is enough to see that $\boldsymbol{e}_1 L$ and $\boldsymbol{e}_2 L$ are contained in L, but these assertions are a direct consequence of the orthonormal relations.

Since we wish that $\mathcal{G}_2$ be a minimal solution to the problem of multiplying vectors, it is natural to assume, as we will do henceforth, that $\langle 1, \boldsymbol{e}_1, \boldsymbol{e}_2, \boldsymbol{e}_1\boldsymbol{e}_2 \rangle = \mathcal{G}_2$.

1.2.3 (Linear basis of $\mathcal{G}_2$) *The elements* $1, \boldsymbol{e}_1, \boldsymbol{e}_2, \boldsymbol{e}_1\boldsymbol{e}_2 \in \mathcal{G}_2$ *are linearly independent. In particular,* $\dim \mathcal{G}_2 = 4$. *Moreover, the linear subspaces* $\mathcal{G}_2^0 = \langle 1 \rangle = \mathbb{R}$, $\mathcal{G}_2^1 = \langle \boldsymbol{e}_1, \boldsymbol{e}_2 \rangle = E_2$, *and* $\mathcal{G}_2^2 = \langle \boldsymbol{e}_1\boldsymbol{e}_2 \rangle$ *are independent of the orthonormal basis used to construct them.*

Proof Assume a linear relation of the form

$$\lambda + \lambda_1 \boldsymbol{e}_1 + \lambda_2 \boldsymbol{e}_2 + \lambda_{12} \boldsymbol{e}_1 \boldsymbol{e}_2 = 0.$$

Multiplying by $\boldsymbol{e}_1$ from the left and from the right, we get the relation

$$\lambda + \lambda_1 \boldsymbol{e}_1 - \lambda_2 \boldsymbol{e}_2 - \lambda_{12} \boldsymbol{e}_1 \boldsymbol{e}_2 = 0.$$

Adding the last two relations, we conclude that $\lambda + \lambda_1 \boldsymbol{e}_1 = 0$ and hence $\lambda = \lambda_1 = 0$. So we are left with the relation $\lambda_2 \boldsymbol{e}_2 + \lambda_{12} \boldsymbol{e}_1 \boldsymbol{e}_2 = 0$ or, multiplying on the right by $\boldsymbol{e}_2$, $\lambda_2 + \lambda_{12} \boldsymbol{e}_1 = 0$, which gives $\lambda_2 = \lambda_{12} = 0$.

Since the assertions on $\mathcal{G}_2^0 = \mathbb{R}$ and $\mathcal{G}_2^1 = E_2$ are clear, what remains is to see that $\mathcal{G}_2^2$ is independent of the basis. We will do this by giving a description of it that is basis independent.

Consider the map $E_2^2 \to \mathcal{G}_2$ given by $(\boldsymbol{v}, \boldsymbol{w}) \mapsto \frac{1}{2}(\boldsymbol{v}\boldsymbol{w} - \boldsymbol{w}\boldsymbol{v})$. Since this map is bilinear and skew-symmetric, it gives a linear map $\wedge^2 E_2 \to \mathcal{G}_2$ (which is basis independent by definition) such that $\boldsymbol{v} \wedge \boldsymbol{w} \mapsto \frac{1}{2}(\boldsymbol{v}\boldsymbol{w} - \boldsymbol{w}\boldsymbol{v})$. In particular,

$$\boldsymbol{e}_1 \wedge \boldsymbol{e}_2 \mapsto \tfrac{1}{2}(\boldsymbol{e}_1\boldsymbol{e}_2 - \boldsymbol{e}_2\boldsymbol{e}_1) = \boldsymbol{e}_1\boldsymbol{e}_2,$$

which shows that the map in question yields a canonical linear isomorphism

$$\wedge^2 E_2 \simeq \mathcal{G}_2^2.$$

□

Thus we actually have a canonical linear isomorphism

$$\wedge E_2 = \wedge^0 E_2 \oplus \wedge^1 E_2 \oplus \wedge^2 E_2 \simeq \mathcal{G}_2^0 + \mathcal{G}_2^1 + \mathcal{G}_2^2 = \mathcal{G}_2.$$

This isomorphism shows that $\mathcal{G}_2$ is endowed with a natural grading as a vector space and allows us to copy all the features and terminology of the exterior algebra $\wedge E_2$ onto $\mathcal{G}_2$. In particular, we have the *exterior product* (or *outer product*) in $\mathcal{G}_2$ coexisting, from now on, with the geometric product. The basic relation between the two products is the following formula, also discovered by Clifford [from now on we will use the expression $\boldsymbol{v} \cdot \boldsymbol{w}$ as an alternative notation for $q(\boldsymbol{v}, \boldsymbol{w})$]:

1.2.4 (Key formula) $\boldsymbol{v}\boldsymbol{w} = \boldsymbol{v} \cdot \boldsymbol{w} + \boldsymbol{v} \wedge \boldsymbol{w}$.

Proof Indeed, $\boldsymbol{v}\boldsymbol{w} = \frac{1}{2}(\boldsymbol{v}\boldsymbol{w} + \boldsymbol{w}\boldsymbol{v}) + \frac{1}{2}(\boldsymbol{v}\boldsymbol{w} - \boldsymbol{w}\boldsymbol{v}) = \boldsymbol{v} \cdot \boldsymbol{w} + \boldsymbol{v} \wedge \boldsymbol{w}$. □

1.2.5 (Involutions of $\mathcal{G}_2$) *The parity and reverse involutions of* $\wedge E_2$ *are also involutions of the geometric product:* $\widehat{xy} = \hat{x}\,\hat{y}$ *and* $\widetilde{xy} = \tilde{y}\,\tilde{x}$.

Proof From the definitions it follows that we may assume that x and y are homogeneous. Let k and l be their grades. If $k = l = 1$, the claims follow from the key formula (using the analogous rules for the exterior product). If $k = l = 2$, it suffices to consider the case $x = y = \boldsymbol{e}_1\boldsymbol{e}_2$ and this is immediate because $\boldsymbol{e}_1\boldsymbol{e}_2\boldsymbol{e}_1\boldsymbol{e}_2 = -1$. Finally, the mixed case follows from the fact that $\boldsymbol{e}_1\boldsymbol{e}_2$ anticommutes with vectors. □

As we have seen, the area element $\boldsymbol{i} = \boldsymbol{e}_1\boldsymbol{e}_2$ is defined up to sign by E_2. The two elements $\pm\boldsymbol{i}$ are called *unit areas* or *orientations* of E_2, and E_2 is considered to be *oriented* when one of the two unit areas is chosen as the positive orientation. Henceforth we will assume that E_2 is oriented and we will let $\boldsymbol{i}$ denote the positive unit area. For an arbitrary basis $\boldsymbol{v}_1, \boldsymbol{v}_2$ of E_2, we have that $\boldsymbol{v}_1 \wedge \boldsymbol{v}_2 = \delta\boldsymbol{i}$, $\delta \in \mathbb{R}$, and we say that that basis is *positive* or *negative* according to whether $\delta > 0$ or $\delta < 0$.

The positive unit area $\boldsymbol{i}$ plays a very important role in what follows. Letting $\boldsymbol{e}_1, \boldsymbol{e}_2$ be a positive orthonormal basis, then $\boldsymbol{i} = \boldsymbol{e}_1 \wedge \boldsymbol{e}_2 = \boldsymbol{e}_1\boldsymbol{e}_2$, and its fundamental property is that $\boldsymbol{i}^2 = \boldsymbol{e}_1\boldsymbol{e}_2\boldsymbol{e}_1\boldsymbol{e}_2 = -\boldsymbol{e}_1^2\boldsymbol{e}_2^2 = -1$, which may be stated by saying the $\boldsymbol{i}$ is a *geometric square root* of -1.

If we set $\mathcal{G}_2^+ = \mathcal{G}_2^0 + \mathcal{G}_2^2 = \langle 1, \boldsymbol{i}\rangle$, which is called the *even geometric algebra*, we see that $\mathcal{G}_2^+ \simeq \mathbb{C}$, via the mapping $\alpha + \beta\boldsymbol{i} \mapsto \alpha + \beta i$. Note, however, that the geometric meaning of $\boldsymbol{i}$ is lost when we map it to i, the *formal* square root of -1. To retain the geometric meaning, we set $\mathbf{C} = \mathcal{G}_2^+$ and say that $\mathbf{C}$ is the field of *complex scalars*.

Remark also that the map $\mathbb{R} = \mathcal{G}_2^0 \to \mathcal{G}_2^2$, $\lambda \mapsto \lambda\boldsymbol{i}$, is a linear isomorphism, with inverse the map $s \mapsto -s\boldsymbol{i}$. Because of this isomorphism, the elements of $\mathcal{G}_2^2$ are also called *pseudoscalars*. Since $\boldsymbol{e}_1\boldsymbol{i} = \boldsymbol{e}_2$ and $\boldsymbol{e}_2\boldsymbol{i} = -\boldsymbol{e}_1$, we see that the map $E_2 \to E_2$, $\boldsymbol{v} \mapsto \boldsymbol{v}\boldsymbol{i}$, is a linear isomorphism, with inverse $\boldsymbol{v} \mapsto -\boldsymbol{v}\boldsymbol{i}$. Thus, in particular, E_2 becomes a $\mathbf{C}$-vector space.

It is a good moment to introduce the *oriented angle* $\alpha_{\boldsymbol{i}} = \alpha_{\boldsymbol{i}}(\boldsymbol{v}, \boldsymbol{v}')$ between two non-zero vectors $\boldsymbol{v}$ and $\boldsymbol{v}'$. Since we impose that this angle does not depend on the lengths of the vectors, we may assume that $\boldsymbol{v}$ and $\boldsymbol{v}'$ are unit vectors. Then $\{\boldsymbol{v}, \boldsymbol{v}\boldsymbol{i}\}$ is a positive orthonormal basis and $\alpha_{\boldsymbol{i}}$ is defined as the unique scalar $\alpha_{\boldsymbol{i}} \in [0, 2\pi)$ such that $\boldsymbol{v}' = \boldsymbol{v}\cos\alpha_{\boldsymbol{i}} + \boldsymbol{v}\boldsymbol{i}\sin\alpha_{\boldsymbol{i}}$ (see Fig. 1.1c). We clearly have $\alpha_{\boldsymbol{i}} = \alpha = 0$ if $\boldsymbol{v}' = \boldsymbol{v}$ and $\alpha_{\boldsymbol{i}} = \alpha = \pi$ if $\boldsymbol{v}' = -\boldsymbol{v}$.

1.2.6 (Meaning of the oriented angle) *We have* $\boldsymbol{v} \wedge \boldsymbol{v}' = \boldsymbol{i} \sin \alpha_{\boldsymbol{i}}$. *In particular we see that if* $\boldsymbol{v}$ *and* $\boldsymbol{v}'$ *are linearly independent, then* $\boldsymbol{v} \wedge \boldsymbol{v}'$ *has positive or negative orientation according to whether* $\alpha_{\boldsymbol{i}} \in (0, \pi)$, *in which case* $\alpha_{\boldsymbol{i}} = \alpha$, *or* $\alpha_{\boldsymbol{i}} \in (\pi, 2\pi)$, *in which case* $\alpha_{\boldsymbol{i}} = 2\pi - \alpha$.

Proof On the one hand we have, by the key formula, $\boldsymbol{v}\boldsymbol{v}' = \cos\alpha + \boldsymbol{v} \wedge \boldsymbol{v}'$. On the other, from the definition of $\alpha_{\boldsymbol{i}}$ we have that $\boldsymbol{v}\boldsymbol{v}' = \cos\alpha_{\boldsymbol{i}} + \boldsymbol{i} \sin\alpha_{\boldsymbol{i}}$. Therefore $\cos\alpha_{\boldsymbol{i}} = \cos\alpha$ and $\boldsymbol{v} \wedge \boldsymbol{v}' = \boldsymbol{i} \sin\alpha_{\boldsymbol{i}}$. The first equality tells us that either $\alpha_{\boldsymbol{i}} = \alpha \in (0, \pi)$ or $\alpha_{\boldsymbol{i}} = 2\pi - \alpha \in (\pi, 2\pi)$. In the first case, $\boldsymbol{v} \wedge \boldsymbol{v}' = \boldsymbol{i} \sin\alpha$ and $\boldsymbol{v} \wedge \boldsymbol{v}'$ is positive. In the second case, $\boldsymbol{v} \wedge \boldsymbol{v}' = \boldsymbol{i} \sin(2\pi - \alpha) = -\boldsymbol{i} \sin\alpha$ and $\boldsymbol{v} \wedge \boldsymbol{v}'$ is negative. □

We define $\mathrm{P}^- \subset \mathcal{G}_2^- = E_2$ to be the set of unit vectors (the circle of radius 1 centered at 0) and $\mathrm{P}^+ \subset \mathcal{G}_2^+ = \mathbf{C}$ to be the set of unit complex scalars (complex scalars z such that $z\tilde{z} = 1$). Since $\tilde{z} = \alpha - \beta\boldsymbol{i}$ if $z = \alpha + \beta\boldsymbol{i}$, $z\tilde{z} = \alpha^2 + \beta^2$ and therefore $\mathrm{P}^+ = \{z_\varphi = e^{\boldsymbol{i}\varphi} : \varphi \in [0, 2\pi)\}$ and

$$\mathrm{P}^- = \{\boldsymbol{u}_\alpha = \boldsymbol{e}_1 \cos\alpha + \boldsymbol{e}_2 \sin\alpha = \boldsymbol{e}_1 z_\alpha : \alpha \in [0, 2\pi)\}.$$

Notice that P^+ is a group with the geometric product (*spinor group of* E_2, also denoted Spin_2).

1.2.7 (Symmetries) (1) *For any non-zero vector* $\boldsymbol{v} \in E_2$, $s_{\boldsymbol{v}} = \underline{\boldsymbol{v}}$, *where* $\underline{\boldsymbol{v}}(\boldsymbol{x}) = \boldsymbol{v}\boldsymbol{x}\boldsymbol{v}^{-1}$. (2) *If* $\boldsymbol{w} \in E_2$ *is non-zero, then* $s_{\boldsymbol{w}} = s_{\boldsymbol{v}}$ *if and only if* $\boldsymbol{w} = \lambda\boldsymbol{v}$ *for some* $\lambda \in \mathbb{R}$. (3) *The map* $\mathrm{P}^- \to \mathrm{O}_2^-$, $\boldsymbol{u} \mapsto s_{\boldsymbol{u}}$, *is onto and* $s_{\boldsymbol{u}'} = s_{\boldsymbol{u}}$ *if and only if* $\boldsymbol{u}' = \pm\boldsymbol{u}$ (the map is 2 to 1).

Proof (1) The expression $\boldsymbol{v}\boldsymbol{x}\boldsymbol{v}^{-1}$ is linear in $\boldsymbol{x}$, its value for $\boldsymbol{x} = \boldsymbol{v}$ is $\boldsymbol{v}$, and for $\boldsymbol{x} = \boldsymbol{v}' \in \boldsymbol{v}^\perp$ it is $-\boldsymbol{v}'$, because $\boldsymbol{v}\boldsymbol{v}' = -\boldsymbol{v}'\boldsymbol{v}$. This proves the first assertion.

(2) If $\boldsymbol{w} = \lambda\boldsymbol{v}$, then it is clear that $\boldsymbol{w}\boldsymbol{x}\boldsymbol{w}^{-1} = \boldsymbol{v}\boldsymbol{x}\boldsymbol{v}^{-1}$ for any $\boldsymbol{x} \in E_2$ and hence $s_{\lambda\boldsymbol{v}} = s_{\boldsymbol{v}}$. Conversely, if $s_{\boldsymbol{w}} = s_{\boldsymbol{v}}$, then in particular

$$\boldsymbol{w}\boldsymbol{v}\boldsymbol{w}^{-1} = s_{\boldsymbol{w}}(\boldsymbol{v}) = s_{\boldsymbol{v}}(\boldsymbol{v}) = \boldsymbol{v}.$$

But this relation says that $\boldsymbol{w}\boldsymbol{v} = \boldsymbol{v}\boldsymbol{w}$, hence $\boldsymbol{w} \wedge \boldsymbol{v} = 0$, and this yields our claim.

(3) By definition, the elements of O_2^- are the symmetries and by (2) any symmetry has the form $s_{\boldsymbol{u}}$, with $\boldsymbol{u} \in \mathrm{P}^-$. So the stated map is onto. Finally note that if $\boldsymbol{u}, \boldsymbol{u}' \in \mathrm{P}^-$ and $s_{\boldsymbol{u}} = s_{\boldsymbol{u}'}$, then $\boldsymbol{u}' = \pm\boldsymbol{u}$. □

To get the corresponding result for rotations, first note that the previous result and the definition of rotation imply that any $f \in \mathrm{O}_2^+$ has the form $s_{\boldsymbol{u}'}s_{\boldsymbol{u}}$ with $\boldsymbol{u}, \boldsymbol{u}' \in \mathrm{P}^-$.

1.2.8 (Rotations) (1) *Given* $\boldsymbol{u}, \boldsymbol{u}' \in \mathrm{P}^-$, $z = \boldsymbol{u}'\boldsymbol{u} \in \mathrm{P}^+$. (2) $z = e^{-\boldsymbol{i}\varphi}$, *where* $\varphi = \alpha_{\boldsymbol{i}}(\boldsymbol{u}, \boldsymbol{u}')$. (3) *If* $f = s_{\boldsymbol{u}'}s_{\boldsymbol{u}}$, *then* $f = \underline{z}$, *where* $\underline{z}(\boldsymbol{x}) = z\boldsymbol{x}\tilde{z}$. (4) $\underline{z}(\boldsymbol{x}) = \boldsymbol{x}e^{2\varphi\boldsymbol{i}}$ (see Fig. 1.2a). (5) O_2^+ *is a subgroup of* O_2 *and the map* $\mathrm{P}^+ \to \mathrm{O}_2^+$, $z \mapsto \underline{z}$, *is an onto* 2 *to* 1 *group homomorphism.*

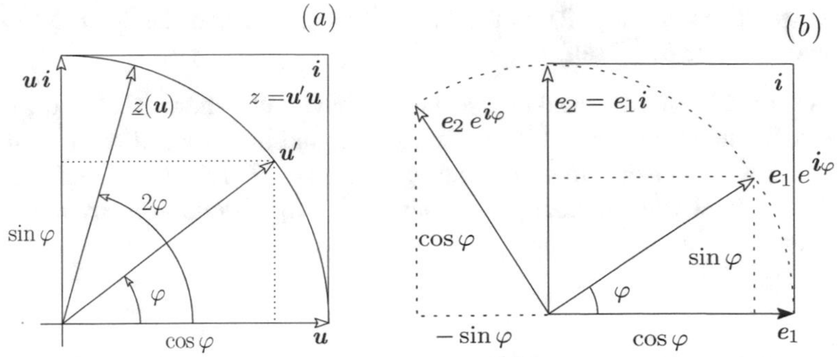

Fig. 1.2 In both graphics $\boldsymbol{i}$ is the unit area. (**a**) Action of the map $\underline{z}$ defined in 1.2.8 on the vector $\boldsymbol{u}$. (**b**) Action of $e^{\varphi \boldsymbol{i}}$ on $\boldsymbol{e}_1$ and $\boldsymbol{e}_2$ (it is the counterclockwise rotation by φ)

Proof (1) Indeed, $z\tilde{z} = \boldsymbol{u}'\boldsymbol{u}\boldsymbol{u}\boldsymbol{u}' = 1$.
(2) Indeed, $\boldsymbol{u}'\boldsymbol{u} = \boldsymbol{u}' \cdot \boldsymbol{u} - \boldsymbol{u} \wedge \boldsymbol{u}' = \cos\varphi - \boldsymbol{i}\sin\varphi = e^{-\varphi \boldsymbol{i}}$.
(3) We have $f(\boldsymbol{x}) = s_{\boldsymbol{u}'}(s_{\boldsymbol{u}}(\boldsymbol{x})) = \boldsymbol{u}'(\boldsymbol{u}\boldsymbol{x}\boldsymbol{u})\boldsymbol{u}' = z\boldsymbol{x}\tilde{z} = \underline{z}(\boldsymbol{x})$.
(4) Since $\boldsymbol{i}$ anticommutes with vectors, $\underline{z}(\boldsymbol{x}) = z\boldsymbol{x}\tilde{z} = \boldsymbol{x}\tilde{z}^2 = \boldsymbol{x}e^{2\varphi \boldsymbol{i}}$.
(5) First let us check that the map, which is clearly onto, satisfies $\underline{zz'} = \underline{z}\,\underline{z}'$ for $z, z' \in \mathrm{P}^+$: $\underline{zz'}(\boldsymbol{x}) = (zz')\boldsymbol{x}(\widetilde{zz'}) = zz'\boldsymbol{x}\tilde{z}'\tilde{z} = \underline{z}(\underline{z}'(\boldsymbol{x}))$. This implies that O_2^+ is closed under composition and therefore that it is a subgroup of O_2. Finally, if $z = e^{\varphi \boldsymbol{i}}$, $z' = e^{\varphi' \boldsymbol{i}}$ and $\underline{z} = \underline{z}'$, then in particular we have $\boldsymbol{e}_1 e^{2\varphi \boldsymbol{i}} = \boldsymbol{e}_1 e^{2\varphi' \boldsymbol{i}}$, which implies that $e^{2(\varphi'-\varphi)\boldsymbol{i}} = 1$ and hence that $\varphi' = \varphi$ or $\varphi' = \varphi + \pi$. □

The results on symmetries and rotations can be combined in a more comprehensive presentation of O_2.

1.2.9 (GA structure of O_2) (1) *The set* $\mathrm{P} = \mathrm{P}^- \sqcup \mathrm{P}^+ \subset \mathcal{G}_2$ *is a group with the geometric product* (pinor group of E_2, also denoted Pin_2). (2) *The map* $\mathrm{P} \to \mathrm{O}_2$ *obtained combining the maps* $\mathrm{P}^- \to \mathrm{O}_2^-$ *and* $\mathrm{P}^+ \to \mathrm{O}_2^+$ *is given by* $p \mapsto \underline{p}$, *where* $\underline{p}(\boldsymbol{x}) = p\boldsymbol{x}\tilde{p}$ (spinorial map). (3) *The map* $\mathrm{P} \to \mathrm{O}_2$ *is a 2 to 1 and onto group homomorphism.*

Proof (1) It is enough to check that P is closed under the geometric product. We have $\mathrm{P}^+\mathrm{P}^+ \subset \mathrm{P}^+$ (for P^+ is a group), $\mathrm{P}^-\mathrm{P}^- \subset \mathrm{P}^+$ [1.2.8 (1)], and $\mathrm{P}^-\mathrm{P}^+, \mathrm{P}^+\mathrm{P}^- \subset \mathrm{P}^-$ (these follow from 1.2.8 (4) and the fact that $z\boldsymbol{u} = \boldsymbol{u}\tilde{z}$ for $z \in \mathrm{P}^+$ and $\boldsymbol{u} \in \mathrm{P}^-$).
(2) By 1.2.7 and 1.2.8 (3).
(3) It is immediate to check that it is a group homomorphism [it goes like the proof of 1.2.8 (5)] and all other claims are obvious. □

1.2.10 (Remark) The Euclidean plane E_2 is special with respect to the general theory in two important aspects: The group P^+ is commutative and is *isomorphic* (sic) to the rotation group O_2^+, for the rotation by an angle φ is given by $\boldsymbol{x} \mapsto \boldsymbol{x}e^{\varphi \boldsymbol{i}}$ (see Fig. 1.2b). But the *spinorial map* $\mathrm{P}^+ \to \mathrm{O}_2^+$, seen as a map $\mathrm{P}^+ \to \mathrm{P}^+$, is $z \mapsto z^2$. Topologically, P^+ is the unit 1-sphere, S^1, and the spinorial map is wrapping S^1 twice over itself.

1.2.11 (Remark) Strictly speaking, so far we have only shown that if $\mathcal{G}_2$ exists, then it is unique up to a canonical isomorphism, and that it has dimension 4. The existence can be shown using the Example 1.1.3 (see E.1.6, p. 32), but it can be proved in general without resorting to matrices. This is done in many references, and in particular in [97].

1.3 GA of the Euclidean Space, $\mathcal{G}_3$

The aim of this section is to explore the geometric algebra $\mathcal{G}_3$ of the Euclidean space E_3 and its bearing on the geometry of this space.

Geometric Background

Let us first recall the Cartan-Dieudonné theorem for the isometry group O_3 of E_3. The notion of symmetry $s_{\boldsymbol{v}}$ can also be defined for E_3 (and for any E_n) in exactly the same way as for E_2. For $n > 2$, however, we actually need the notion of *reflection* $r_{\boldsymbol{v}} \in O_3$ associated with a non-zero vector $\boldsymbol{v}$, which is defined as $-s_{\boldsymbol{v}}$. In other words, $r_{\boldsymbol{v}}$ is the unique linear map $E_3 \to E_3$ such that $r_{\boldsymbol{v}}(\boldsymbol{v}) = -\boldsymbol{v}$ and $r_{\boldsymbol{v}}(\boldsymbol{x}) = \boldsymbol{x}$ for $\boldsymbol{x} \in \boldsymbol{v}^{\perp}$. The reason why we did not speak of reflections in the case $n = 2$ is that for this dimension we have $s_{\boldsymbol{v}} = r_{\boldsymbol{v}'}$ for any non-zero vector $\boldsymbol{v}'$ orthogonal to $\boldsymbol{v}$, which tells us that for E_2 the use of symmetries is equivalent to the use of reflections.

1.3.1 (Cartan-Dieudonné theorem for E_3) *Any isometry of E_3 is the composition of at most three reflections.*

Proof Let $f \in O_3$. If $f(\boldsymbol{x}) = \boldsymbol{x}$ for all $\boldsymbol{x}$, f is the identity, which is the composition of zero reflections (or the square of any reflection). So we may assume that there is a vector $\boldsymbol{v}$ such that $\boldsymbol{w} = f(\boldsymbol{v}) - \boldsymbol{v} \neq 0$. Then we have $r_{\boldsymbol{w}}(f(\boldsymbol{v}) - \boldsymbol{v}) = -f(\boldsymbol{v}) + \boldsymbol{v}$ and, since $f(\boldsymbol{v}) + \boldsymbol{v}$ is orthogonal to $\boldsymbol{w}$, $r_{\boldsymbol{w}}(f(\boldsymbol{v}) + \boldsymbol{v}) = f(\boldsymbol{v}) + \boldsymbol{v}$. It follows that $r_{\boldsymbol{w}}(f(\boldsymbol{v})) = \boldsymbol{v}$. This means that it will be enough to show that an isometry leaving a vector fixed is the composition of at most two reflections. But this follows easily by considering the isometry f' of $\boldsymbol{v}^{\perp}$ induced by f and the fact that f' is the composition of at most two reflections. □

The Cartan-Dieudonné theorem shows that $O_3 = O_3^+ \sqcup O_3^-$, where O_3^+ is formed by the isometries that are the composition of two reflections (which includes the identity) and O_3^- by those that are either a reflection or the composition of three reflections. The set O_3^+ is also denoted SO_3 and its elements are called *rotations*.

Geometric Algebra

The initial steps are similar to those followed for $\mathcal{G}_2$ and so it will suffice to recall the main points. We are looking for an algebra $\mathcal{G}_3$ that contains E_3, with $\mathbb{R} \cap E_3 = \{0\}$ and such that $\boldsymbol{v}^2 = q(\boldsymbol{v})$ for any vector $\boldsymbol{v}$, or, equivalently, such that

$$\boldsymbol{vw} + \boldsymbol{wv} = 2q(\boldsymbol{v}, \boldsymbol{w})$$

for all $\boldsymbol{v}, \boldsymbol{w} \in E_3$. In particular, $\boldsymbol{vw} = -\boldsymbol{wv}$ if and only if $\boldsymbol{v}$ and $\boldsymbol{w}$ are orthogonal.

Let $\boldsymbol{e}_1, \boldsymbol{e}_2, \boldsymbol{e}_3$ be any orthonormal basis of E_3. Define $B_0 = \{1\}$, $B_1 = \{\boldsymbol{e}_1, \boldsymbol{e}_2, \boldsymbol{e}_3\}$, $B_2 = \{\boldsymbol{e}_2\boldsymbol{e}_3, \boldsymbol{e}_3\boldsymbol{e}_1, \boldsymbol{e}_1\boldsymbol{e}_2\}$, $B_3 = \{\boldsymbol{e}_1\boldsymbol{e}_2\boldsymbol{e}_3\}$ and $B = B_0 \cup B_1 \cup B_2 \cup B_3$. The elements of B are called *Clifford units* (associated with the orthonormal basis $\boldsymbol{e}_1, \boldsymbol{e}_2, \boldsymbol{e}_3$).

1.3.2 (Clifford bases) *B is linearly independent.*

Proof We will adapt the method used for E_2 (Riesz' method, [82]) to the present case. Suppose we have a linear relation

$$\lambda + \lambda_1\boldsymbol{e}_1 + \lambda_2\boldsymbol{e}_2 + \lambda_3\boldsymbol{e}_3 + \mu_1\boldsymbol{e}_2\boldsymbol{e}_3 + \mu_2\boldsymbol{e}_3\boldsymbol{e}_1 + \mu_3\boldsymbol{e}_1\boldsymbol{e}_2 + \mu\boldsymbol{e}_1\boldsymbol{e}_2\boldsymbol{e}_3 = 0.$$

Multiplying by $\boldsymbol{e}_1$ from the left and from the right, we get

$$\lambda + \lambda_1\boldsymbol{e}_1 - \lambda_2\boldsymbol{e}_2 - \lambda_3\boldsymbol{e}_3 + \mu_1\boldsymbol{e}_2\boldsymbol{e}_3 - \mu_2\boldsymbol{e}_3\boldsymbol{e}_1 - \mu_3\boldsymbol{e}_1\boldsymbol{e}_2 + \mu\boldsymbol{e}_1\boldsymbol{e}_2\boldsymbol{e}_3 = 0.$$

Adding the two equations, we arrive at

$$\lambda + \lambda_1\boldsymbol{e}_1 + \mu_1\boldsymbol{e}_2\boldsymbol{e}_3 + \mu\boldsymbol{e}_1\boldsymbol{e}_2\boldsymbol{e}_3 = 0.$$

Now multiply by $\boldsymbol{e}_2$ from the left and from the right. This yields

$$\lambda - \lambda_1\boldsymbol{e}_1 - \mu_1\boldsymbol{e}_2\boldsymbol{e}_3 + \mu\boldsymbol{e}_1\boldsymbol{e}_2\boldsymbol{e}_3 = 0,$$

and hence $\lambda + \mu\boldsymbol{e}_1\boldsymbol{e}_2\boldsymbol{e}_3 = 0$. But $(\boldsymbol{e}_1\boldsymbol{e}_2\boldsymbol{e}_3)^2 = -1$ and so $\lambda^2 = -\mu^2$, which implies $\lambda = \mu = 0$. Thus $\lambda_1\boldsymbol{e}_1 + \mu_1\boldsymbol{e}_2\boldsymbol{e}_3 = 0$. Multiplying by $\boldsymbol{e}_1$ from the left, we arrive at an equation that allows us to conclude, as in the previous step, that $\lambda_1 = \mu_1 = 0$. So we are left with the relation

$$\lambda_2\boldsymbol{e}_2 + \lambda_3\boldsymbol{e}_3 + \mu_2\boldsymbol{e}_3\boldsymbol{e}_1 + \mu_3\boldsymbol{e}_1\boldsymbol{e}_2 = 0.$$

To conclude we can repeat the game with this equation. Multiplying by $\boldsymbol{e}_2$ from the left and from the right, we obtain $\lambda_2 = \mu_2 = 0$, and then $\lambda_3 = \mu_3 = 0$ follows readily. □

With a similar argument as the one used for E_2, we see that any product of vectors is a linear combination of B and hence we conclude that

$$\mathcal{G}_3 = \langle B \rangle, \tag{1.6}$$

because we assume that $\mathcal{G}_3$ is generated, as an $\mathbb{R}$-algebra, by E_3. In particular we have $\dim \mathcal{G}_3 = 8$.

1.3.3 (Linear grading of $\mathcal{G}_3$) *There is a canonical linear map $\wedge^k E_3 \to \mathcal{G}_3$ whose image is $\mathcal{G}_3^k = \langle B_k \rangle$. It follows that the linear map $\wedge^k E_3 \to \mathcal{G}_3^k$ is an isomorphism and in particular that the spaces $\mathcal{G}_3^k$ are independent of the basis $\boldsymbol{e}_1, \boldsymbol{e}_2, \boldsymbol{e}_3$ used to define them.*

Proof Since $\wedge^0 E_3 = \mathbb{R} = \langle 1 \rangle = \mathcal{G}_3^0$ and $\wedge^1 E_3 = E_3 = \langle B_1 \rangle = \mathcal{G}_3^1$, we only need to consider the cases $k = 2, 3$.

The map $\wedge^2 E_3 \to \mathcal{G}_3^2$ is the unique linear map such that $\boldsymbol{v} \wedge \boldsymbol{w} \mapsto \frac{1}{2}(\boldsymbol{v}\boldsymbol{w} - \boldsymbol{w}\boldsymbol{v})$. Its image is $\mathcal{G}_3^2$, as $\boldsymbol{e}_i \wedge \boldsymbol{e}_j \mapsto \frac{1}{2}(\boldsymbol{e}_i\boldsymbol{e}_j - \boldsymbol{e}_j\boldsymbol{e}_i) = \boldsymbol{e}_i\boldsymbol{e}_j$ and hence $\wedge^2 E_3 \simeq \mathcal{G}_3^2$.

The map $\wedge^3 E_3 \to \mathcal{G}_3^3$ is the unique linear map such that

$$\boldsymbol{u} \wedge \boldsymbol{v} \wedge \boldsymbol{w} \mapsto \tfrac{1}{6}(\boldsymbol{u}\boldsymbol{v}\boldsymbol{w} + \boldsymbol{v}\boldsymbol{w}\boldsymbol{u} + \boldsymbol{w}\boldsymbol{u}\boldsymbol{v} - \boldsymbol{u}\boldsymbol{w}\boldsymbol{v} - \boldsymbol{v}\boldsymbol{u}\boldsymbol{w} - \boldsymbol{w}\boldsymbol{v}\boldsymbol{u}).$$

Note that the right-hand side is the full anti-symmetrization of the product $\boldsymbol{u}\boldsymbol{v}\boldsymbol{w}$ and hence it is 3-multilinear and skew-symmetric expression of $\boldsymbol{u}, \boldsymbol{v}, \boldsymbol{w}$. It follows that $\boldsymbol{e}_1 \wedge \boldsymbol{e}_2 \wedge \boldsymbol{e}_3 \mapsto \boldsymbol{e}_1\boldsymbol{e}_2\boldsymbol{e}_3$ and hence $\wedge^3 E_3 \simeq \mathcal{G}_3^3$. □

So we have a canonical **linear** isomorphism $\wedge E_3 \simeq \mathcal{G}_3$. With it we can transfer to $\mathcal{G}_3$ the concepts pertaining to $\wedge E_3$. Thus we will say that the elements of $\mathcal{G}_3$ are *multivectors*, with suitable names for k-vectors for the various k (see the left column in Fig. 1.3 for a synopsis).

Scalars	1
Vectors –oriented segments –*polar vectors*	$\mathbf{e}_1, \mathbf{e}_2, \mathbf{e}_3$
Bivectors –oriented areas –*axial vectors*	$\mathbf{e}_1\mathbf{i}, \mathbf{e}_2\mathbf{i}, \mathbf{e}_3\mathbf{i}$
Pseudoscalars –*oriented volumes*	$\mathbf{i} = \mathbf{e}_1\mathbf{e}_2\mathbf{e}_3$

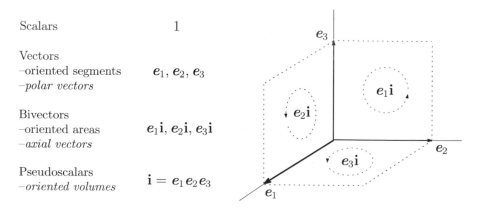

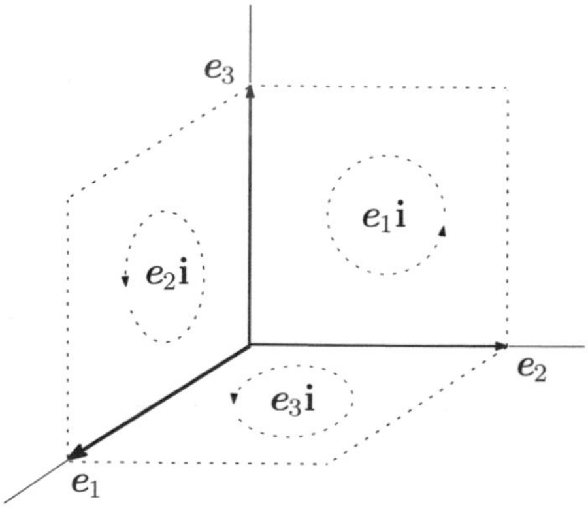

Fig. 1.3 Clifford units of E_3 and Hodge duality

As in the case of $\mathcal{G}_2$, we get an *exterior product* $x \wedge y$ in $\mathcal{G}_3$ (also called *outer product*), which is associative, unital, and skew-commutative. Its relation to the geometric product is ruled by the *key formula* 1.2.4 (same proof) for the product of two vectors. In particular we have $\boldsymbol{e}_j \wedge \boldsymbol{e}_k = \boldsymbol{e}_j \boldsymbol{e}_k$. Note also that $\wedge^3 E_3 \simeq \mathcal{G}_3^3$ gives that $\boldsymbol{e}_1 \wedge \boldsymbol{e}_2 \wedge \boldsymbol{e}_3 = \boldsymbol{e}_1 \boldsymbol{e}_2 \boldsymbol{e}_3$.

Like for E_2, the *orientations* of E_3 are $\pm\mathbf{i}$, where $\mathbf{i} = \boldsymbol{e}_1 \boldsymbol{e}_2 \boldsymbol{e}_3$. Henceforth we will assume that E_3 is oriented by $\mathbf{i}$, which is called (positive) *volume element* or *pseudoscalar*. For an arbitrary basis $\boldsymbol{v}_1, \boldsymbol{v}_2, \boldsymbol{v}_3$, $\boldsymbol{v}_1 \wedge \boldsymbol{v}_2 \wedge \boldsymbol{v}_3 = \delta\,\mathbf{i}$ ($\delta \in \mathbb{R}$) and we say that the basis is *positive* or *negative* according to whether $\delta > 0$ or $\delta < 0$. The choice of $\mathbf{i}$, for example, tells us that the orthonormal basis $\boldsymbol{e}_1, \boldsymbol{e}_2, \boldsymbol{e}_3$ used to express $\mathbf{i}$ is positive.

1.3.4 (Properties of $\mathbf{i}$) (1) $\mathbf{i}^2 = -1$. (2) $\mathbf{i}$ *commutes with all vectors and hence commutes with any multivector* (we say that $\mathbf{i}$ is in the center of $\mathcal{G}_3$). (3) *If* $x \in \mathcal{G}_3^k$, *then* $x\mathbf{i} = \mathbf{i}x \in \mathcal{G}_3^{3-k}$ *and the linear map* $\mathcal{G}_3^k \to \mathcal{G}_3^{3-k}$, $x \mapsto x^* = x\mathbf{i}$ *is an isomorphism* (Hodge duality). *The inverse map is* $y \mapsto -y\mathbf{i}$.

Proof (1) $\mathbf{i}^2 = \boldsymbol{e}_1\boldsymbol{e}_2\boldsymbol{e}_3\boldsymbol{e}_1\boldsymbol{e}_2\boldsymbol{e}_3 = \boldsymbol{e}_2\boldsymbol{e}_3\boldsymbol{e}_1\boldsymbol{e}_2 = -1$.

(2) It suffices to see that $\mathbf{i}$ commutes with the basis vectors, and this is checked immediately. For example,

$$\boldsymbol{e}_2\mathbf{i} = \boldsymbol{e}_2\boldsymbol{e}_1\boldsymbol{e}_2\boldsymbol{e}_3 = -\boldsymbol{e}_1\boldsymbol{e}_3 \text{ and } \mathbf{i}\boldsymbol{e}_2 = \boldsymbol{e}_1\boldsymbol{e}_2\boldsymbol{e}_3\boldsymbol{e}_2 = -\boldsymbol{e}_1\boldsymbol{e}_3.$$

(3) Since $\mathbf{i}$ contains the three basis vectors, $\boldsymbol{e}_k\mathbf{i}$ has grade 2 and $\boldsymbol{e}_j\boldsymbol{e}_k\mathbf{i}$ has grade 1. The last statements are now a direct consequence of (1). □

We can also transfer to $\mathcal{G}_3$ the parity and reverse involutions.

1.3.5 (Involutions of $\mathcal{G}_3$) *The parity and reverse involutions of* $\wedge E_3$ *are also involutions of the geometric product:* $\widehat{xy} = \hat{x}\,\hat{y}$ *and* $\widetilde{xy} = \tilde{y}\,\tilde{x}$.

Proof On account of bilinearity, we can assume that x and y are Clifford units, say $x = e_J$ of grade j and $y = e_K$ of grade k. In this case the grade of xy is $j + k - 2\nu$, where $\nu = |J \cap K|$. Thus we have $\hat{x}\,\hat{y} = (-1)^{j+k}xy$ and $\widehat{xy} = (-1)^{j+k-2\nu}xy$, which coincide because the signs are the same. Similarly, $\tilde{y}\,\tilde{x}$ is equal to $(-1)^{k/\!\!/2+j/\!\!/2}yx$, while $\widetilde{xy} = (-1)^{(j+k)/\!\!/2-\nu}xy$. But $yx = (-1)^{kj-\nu}xy$ and so it suffices to show that $k/\!\!/2 + j/\!\!/2 + kj$ and $(j+k)/\!\!/2$ have the same parity. This is easily checked by cases. For example, if both j and k are odd, say $j = 2j' + 1$ and $k = 2k' + 1$, then the first expression has the same parity as $j' + k' + 1$ while $(j+k)/\!\!/2 = (2j' + 2k' + 2)/\!\!/2 = j' + k' + 1$. □

Now we are ready to explore how to express geometric facts about E_3 by means of the algebra $\mathcal{G}_3$.

1.3.6 (Geometric meaning of Hodge duality) *If* $\boldsymbol{u} \in \mathcal{G}_3^1$ *is a unit vector,* $\boldsymbol{a} = \boldsymbol{u}\mathbf{i} \in \mathcal{G}_3^2$ *is the* (oriented) *unit area of the plane* $\boldsymbol{u}^\perp$. *Conversely, if* $\boldsymbol{a} \in \mathcal{G}_3^2$ *is a unit area, then* $\boldsymbol{u} = -\boldsymbol{a}\mathbf{i}$ *is the* (oriented) *orthogonal unit vector to the plane*

$$|a\rangle = \{x \in E_3 \mid a \wedge x = 0\}.$$

Proof The oriented unit area of $u^{\perp}$ is, by definition, $a = u_1 u_2$, where u_1, u_2 is an orthonormal basis of $u^{\perp}$ such that $u_1 u_2 u = \mathbf{i}$. Since $u^2 = 1$, it is obvious that $a = u\mathbf{i}$ and, conversely, that $u = -a\mathbf{i}$. □

1.3.7 (Spinorial formula for reflections) *Let $v \in E_3$ be a unit vector. Then*

$$r_v(x) = -vxv = \hat{v}xv.$$

Proof The expression $\hat{v}xv$ is linear in x. For $x = v$, its value is $\hat{v} = -v$, and if $x \in v^{\perp}$, then $\hat{v}xv = -\hat{v}vx = x$. □

1.3.8 (Elements of a rotation) If f is a rotation, then $f(x) = r_{v'}(r_v(x))$, where v and v' are unit vectors (see the last paragraph of the background subsection at the beginning of this section, page 14). By the preceding proposition, we can write $f(x) = -v'(-vxv)v' = Rx\tilde{R}$, with $R = v'v$. If v and v' are linearly dependent, then $R = \pm 1$ and f is the identity. So we can assume that v and v' are linearly independent. Now the key formula allows us to write $v'v = v \cdot v' - v \wedge v' = \cos\alpha - v \wedge v'$, with $\alpha = \alpha(v, v')$. On the other hand, if we let i be the unit area of the oriented plane $\langle v, v'\rangle$, then $v \wedge v' = i \sin\beta$, where $\beta \in [0, 2\pi)$ is the oriented angle $\alpha_i(v, v')$. But $i = u\mathbf{i}$, where u is the oriented unit vector in $\langle v, v'\rangle^{\perp} = v^{\perp} \cap v'^{\perp}$, and so we finally have $R = \cos\beta - u\mathbf{i}\sin\beta = e^{-\beta u\mathbf{i}}$. Since u anticommutes with v and v', u commutes with R and hence it is fixed by f. We will say that u is the *axis* of the rotation f. We will also say that R is the *rotor* associated with the vectors v and v' and we will denote it by $R_{u,\beta}$. The final touch in the geometric algebra description of f is provided by next proposition.

1.3.9 (Euler's spinorial formula for rotations) With the same notations as in the preceding paragraph, we have $f(x) = R_{u,\beta}x\tilde{R}_{u,\beta}$. Since u is fixed by f, it induces an isometry $f^{\perp}$ of the plane $u^{\perp}$, and we have: *$f^{\perp}$ is the* (oriented) *rotation of $u^{\perp}$ by 2β*. In other words, f is the rotation about u by 2β.

Proof If $x \in u^{\perp} = \langle v, v'\rangle$, then x anticommutes with u, $f^{\perp}(x) = f(x) = x\tilde{R}^2_{u,\beta}$, and we know that this rotates x by 2β in the sense of the orientation i (which is the same as the orientation $v \wedge v'$) of $u^{\perp}$. □

Euler's spinorial form of rotations tells us that the rotation $\rho_{u,\theta}$ about the unit vector u by θ is given by the rotor $R_{u,\theta/2} = e^{-\theta u\mathbf{i}/2} = \cos(\theta/2) - u\mathbf{i}\sin(\theta/2)$. It is an invertible element in $\mathcal{G}_3^+ = \mathcal{G}_3^0 + \mathcal{G}_3^2$, the *even subalgebra* of $\mathcal{G}_3$. Its inverse is $e^{\theta u\mathbf{i}/2} = \tilde{R}_{u,\theta/2}$, so that $R\tilde{R} = 1$. In the next result we find that $\mathcal{G}_3^+$ is a very familiar algebra (at least from a historical point of view), which we will denote by $\mathbf{H}$ for reasons that will become clear in a moment, and we will determine the form of its elements that satisfy the *rotor condition* $R\tilde{R} = 1$.

1.3.10 (Quaternions) (1) *The elements of* $\mathbf{H}$ *can be written in a unique way in the form $h = \lambda + v\mathbf{i}$, $\lambda \in \mathbb{R}$ and $v \in E_3$.* (2) *$h\tilde{h} = \lambda^2 + v^2 \in \mathbb{R}$ is positive if $h \neq 0$, and hence* $\mathbf{H}$ *is a* (skew) *field ($h^{-1} = \tilde{h}/|h|^2$, where $|h| = \sqrt{h\tilde{h}}$).* (3) $\mathbf{H}$ *is isomorphic* (non-canonically) *to the* (skew) *field* $\mathbb{H}$ *of Hamilton's quaternions.* We will say that $\mathbf{H}$ is the field of geometric quaternions. (4) *The set* $\mathrm{Spin}_3 = \{h \in \mathbf{H} \mid |h| = 1\}$ *is a subgroup of* $\mathbf{H}^*$ (the group of non-zero quaternions). *Its elements are called* spinors *and each spinor* $h \neq \pm 1$ *has the form* $h = \tilde{R}_{\boldsymbol{u},\beta}$ *for some unit vector* $\boldsymbol{u}$ *and scalar* $\beta \in [0, 2\pi)$. (5) *If we define, for* $h \in \mathrm{Spin}_3$, $\underline{h} : E_3 \to E_3$ *by* $\underline{h}\boldsymbol{x} = h\boldsymbol{x}\tilde{h}$, *then* $\underline{h} \in \mathrm{SO}_3$ *and* $\underline{h} = \rho_{\boldsymbol{u},2\beta}$. (6) *The map* $\mathrm{Spin}_3 \to \mathrm{SO}_3$, $h \mapsto \underline{h}$, *is a* 2 to 1 *onto homomorphism.*

Proof (1) An element of $\mathbf{H}$ has the form $\lambda + \lambda_1\boldsymbol{e}_1\mathbf{i} + \lambda_2\boldsymbol{e}_2\mathbf{i} + \lambda_3\boldsymbol{e}_3\mathbf{i}$ and so it has the stated expression with $\boldsymbol{v} = \lambda_1\boldsymbol{e}_1 + \lambda_2\boldsymbol{e}_2 + \lambda_3\boldsymbol{e}_3$.

(2) Since $\tilde{h} = \lambda - \boldsymbol{v}\mathbf{i}$, the expression for $h\tilde{h}$ follows because $-\boldsymbol{v}\mathbf{i}\boldsymbol{v}\mathbf{i} = -\boldsymbol{v}^2\mathbf{i}^2 = \boldsymbol{v}^2$.

(3) Let $\boldsymbol{i}_1 = \boldsymbol{e}_3\mathbf{i}$, $\boldsymbol{i}_2 = \boldsymbol{e}_2\mathbf{i}$, $\boldsymbol{i}_3 = \boldsymbol{e}_1\mathbf{i}$. Then $\mathbf{H} = \langle 1, \boldsymbol{i}_1, \boldsymbol{i}_2, \boldsymbol{i}_3\rangle$ and we have

$$\boldsymbol{i}_1^2 = \boldsymbol{i}_2^2 = \boldsymbol{i}_3^2 = -1,\ \boldsymbol{i}_1\boldsymbol{i}_2 = -\boldsymbol{i}_2\boldsymbol{i}_1 = \boldsymbol{i}_3,\ \boldsymbol{i}_2\boldsymbol{i}_3 = -\boldsymbol{i}_3\boldsymbol{i}_2 = \boldsymbol{i}_1,\ \boldsymbol{i}_3\boldsymbol{i}_1 = -\boldsymbol{i}_1\boldsymbol{i}_3 = \boldsymbol{i}_2,$$

which are (up to notation) Hamilton's celebrated defining relations for $\mathbb{H}$.

(4) The first assertion is immediate. If $h = \lambda + \boldsymbol{v}\mathbf{i}$ is a spinor, then $\lambda^2 + |\boldsymbol{v}|^2 = 1$. This implies that there is a unique $\beta \in [0, 2\pi)$ such that $\lambda = \cos\beta$ and $|\boldsymbol{v}| = \sin\beta$. If $h \neq \pm 1$, then $\boldsymbol{v} \neq 0$ and $h = \cos\beta + \boldsymbol{u}\mathbf{i}\sin\beta$, where $\boldsymbol{u}$ is the normalization of $\boldsymbol{v}$. Therefore $h = \tilde{R}_{\boldsymbol{u},\varphi}$.

(5) Obvious by what we have said so far.

(6) If h and h' are spinors, then $\underline{hh'}\boldsymbol{x} = hh'\boldsymbol{x}\widetilde{hh'} = hh'\boldsymbol{x}\tilde{h}'\tilde{h} = \underline{h}\,\underline{h'}\boldsymbol{x}$. This proves that $h \mapsto \underline{h}$ is a homomorphism and Euler's spinorial formula shows that it is onto. To end the proof it suffices to see that if $h\boldsymbol{x}\tilde{h} = \boldsymbol{x}$ for all vectors $\boldsymbol{x}$, then $h = \pm 1$. Indeed, the condition is equivalent to say that $h\boldsymbol{x} = \boldsymbol{x}h$ for all $\boldsymbol{x}$, and in particular that $h\boldsymbol{e}_k = \boldsymbol{e}_k h$ for $k = 1, 2, 3$. Using the expression $h = \cos\beta + \boldsymbol{u}\mathbf{i}\sin\beta$, these relations imply that $\boldsymbol{e}_k\boldsymbol{u}\sin\beta = \boldsymbol{u}\boldsymbol{e}_k\sin\beta$. Then either $\sin\beta = 0$, in which case $h = \cos\beta = \pm 1$, or $\boldsymbol{e}_k\boldsymbol{u} = \boldsymbol{u}\boldsymbol{e}_k$, which cannot happen. □

1.3.11 (Composing rotations: Olinde Rodrigues formulas, [83]) *Given the rotations* $\rho_{\boldsymbol{u},\theta}$ *and* $\rho_{\boldsymbol{u}',\theta'}$, *the composition* $\rho_{\boldsymbol{u}',\theta'}\rho_{\boldsymbol{u},\theta}$ *is a rotation. If it is not the identity, it has the form* $\rho_{\boldsymbol{u}'',\theta''}$, *and Olinde Rodrigues' formulas yield* $\boldsymbol{u}''$ *and* θ'' *in terms of* $\boldsymbol{u}$, $\boldsymbol{u}'$, θ *and* θ' (with $\alpha = \alpha(\boldsymbol{u}, \boldsymbol{u}')$):

$$\cos\tfrac{\theta''}{2} = \cos\tfrac{\theta}{2}\cos\tfrac{\theta'}{2} - \cos\alpha\sin\tfrac{\theta}{2}\sin\tfrac{\theta'}{2},$$

$$\boldsymbol{u}''\sin\tfrac{\theta''}{2} = \boldsymbol{u}\sin\tfrac{\theta}{2}\cos\tfrac{\theta'}{2} + \boldsymbol{u}'\cos\tfrac{\theta}{2}\sin\tfrac{\theta'}{2} - (\boldsymbol{u}\wedge\boldsymbol{u}')\mathbf{i}\sin\tfrac{\theta}{2}\sin\tfrac{\theta'}{2}.$$

Proof On account of the spinorial form of the rotations, the composition is given by $R\boldsymbol{x}\tilde{R}$, with $R = R_{\boldsymbol{u}',\theta'/2}R_{\boldsymbol{u},\theta/2}$. Computing this product using the expressions

$$R_{\boldsymbol{u},\theta/2} = \cos\tfrac{\theta}{2} - \boldsymbol{u}\mathbf{i}\sin\tfrac{\theta}{2} \quad\text{and}\quad R_{\boldsymbol{u}',\theta'/2} = \cos\tfrac{\theta'}{2} - \boldsymbol{u}'\mathbf{i}\sin\tfrac{\theta'}{2},$$

and equating its scalar and bivector parts to those of $R = \cos\frac{\theta''}{2} - \boldsymbol{u}''\mathbf{i}\sin\frac{\theta''}{2}$, we get the two equations. It is important to note that $\boldsymbol{u}'\boldsymbol{u} = \cos\alpha - \boldsymbol{u}\wedge\boldsymbol{u}'$. □

1.3.12 (Geometric covariance) Euler's spinor formula $R\boldsymbol{x}\tilde{R}$ makes sense when we insert a multivector x instead of the vector $\boldsymbol{x}$. In this way we get a linear map $\mathcal{G}_3 \to \mathcal{G}_3$, $x \mapsto R x \tilde{R}$. This map is an *automorphism,* as shown by the relation $Rxy\tilde{R} = Rx\tilde{R}Ry\tilde{R}$. This capacity of $\mathcal{G}_3$, which we like to call *geometric covariance,* is a general feature of geometric algebra. For a first taste, here go a couple of illustrations.

To ease notation, let us write $x' = Rx\tilde{R}$. For the first illustration, let $\boldsymbol{x}$ and $\boldsymbol{y}$ be vectors. Then we have:

$$(\boldsymbol{x}\wedge\boldsymbol{y})' = \tfrac{1}{2}(\boldsymbol{x}\boldsymbol{y} - \boldsymbol{y}\boldsymbol{x})' = \tfrac{1}{2}(\boldsymbol{x}'\boldsymbol{y}' - \boldsymbol{y}'\boldsymbol{x}') = \boldsymbol{x}'\wedge\boldsymbol{y}'.$$

The term $\boldsymbol{x}'\wedge\boldsymbol{y}'$ constructs the area element defined by the rotated vectors $\boldsymbol{x}'$ *and* $\boldsymbol{y}'$, which is the rotated oriented area. On the other hand, $(\boldsymbol{x}\wedge\boldsymbol{y})'$ is the result of rotating the oriented area as a bivector. For the first interpretation, we need to know $\boldsymbol{x}$ and $\boldsymbol{y}$ separately, but for the second only the area element $\boldsymbol{a} = \boldsymbol{x}\wedge\boldsymbol{y}$ matters, not the particular way in which it has been obtained.

For another instance, suppose that we have a rotor $S = e^{-\alpha\boldsymbol{p}\mathbf{i}/2}$, where $\boldsymbol{p}$ is a unit vector and α a real number. Then geometric covariance allows us to construct $S' = RS\tilde{R}$ and its geometric meaning is easily discovered by looking at the following relation:

$$S' = (\cos\tfrac{\alpha}{2} - \boldsymbol{p}\,\mathbf{i}\sin\tfrac{\alpha}{2})' = \cos\tfrac{\alpha}{2} - \boldsymbol{p}'\mathbf{i}\sin\tfrac{\alpha}{2}.$$

Indeed, the right-hand side is the rotor of the rotation by α about the rotated axis $\boldsymbol{p}'$ of the axis of the rotation defined by S. Said in other words, to rotate the rotation $\rho_{\boldsymbol{p},\alpha}$ to $\rho_{\boldsymbol{p}',\alpha}$ it is enough to rotate the rotor S *to* S', for the rotor S' gives $\rho_{\boldsymbol{p}',\alpha}$. In the first interpretation we operate at the level of vectors, while in the second the operations are carried out directly at the level of bivectors.

For more involved examples, and to better appreciate the fundamental importance of geometric covariance, see Chaps. 4 and 5, and in particular the geometric solution of the inverse kinematics problem developed in Sect. 4.4. □

1.3.13 (The Pauli representation of $\mathcal{G}_3$) We found that $\mathcal{G}_2$ is isomorphic to $\mathbb{R}(2)$, and that this gave a proof of the existence of $\mathcal{G}_2$. There is a similar development in the case of $\mathcal{G}_3$, which turns out to be isomorphic to $\mathbb{C}(2)$. This can also be regarded as a proof of the existence of $\mathcal{G}_3$, while uniqueness has been discussed at length all along this section.

The isomorphism of $\mathcal{G}_3$ with $\mathbb{C}(2)$ is called the *Pauli representation* and works as follows (cf. [78]). Consider the matrices

$$\sigma_0 = I_2 = \begin{pmatrix} 1 & 0 \\ 0 & 1 \end{pmatrix}, \quad \sigma_1 = \begin{pmatrix} 0 & 1 \\ 1 & 0 \end{pmatrix}, \quad \sigma_2 = \begin{pmatrix} 0 & -i \\ i & 0 \end{pmatrix}, \quad \sigma_3 = \begin{pmatrix} 1 & 0 \\ 0 & -1 \end{pmatrix}.$$

They are called the *Pauli matrices* and what interests us here is that they satisfy Clifford's relations:

$$\sigma_j\sigma_k + \sigma_k\sigma_j = 2\delta_{j,k} \quad (\delta_{j,k} = 0 \text{ if } j \neq k, \ = 1 \text{ if } j = k).$$

By straightforward computations, it can be checked that

$$\sigma_0,\ \sigma_1,\ \sigma_2,\ \sigma_3,\ \sigma_1\sigma_2,\ \sigma_1\sigma_3,\ \sigma_2\sigma_3,\ \sigma_1\sigma_2\sigma_3$$

is a linear basis of $\mathbb{C}(2)$. If we now map $\boldsymbol{x} = \lambda_1\boldsymbol{e}_1 + \lambda_2\boldsymbol{e}_2 + \lambda_3\boldsymbol{e}_3 \in E_3$ to the matrix $m(\boldsymbol{x}) = \lambda_1\sigma_1 + \lambda_2\sigma_2 + \lambda_3\sigma_3 \in \langle\sigma_1, \sigma_2, \sigma_3\rangle$, then the Clifford relations yield that $m(\boldsymbol{x})^2 = \lambda_1^2 + \lambda_2^2 + \lambda_3^2 = q(\boldsymbol{x})$, which is to be understood as saying that $\mathbb{C}(2)$ is a model of $\mathcal{G}_3$. Note, however, that the rich geometric structure of $\mathcal{G}_3$ is invisible in $\mathbb{C}(2)$, much as the structure of $\mathcal{G}_2$ is invisible in the model $\mathbb{R}(2)$. □

Inner Product and Applications to Vector Algebra

We end the section with a few interesting questions that did not find a proper place in the exposition so far.

1.3.14 (Inner product) The inner product $x \cdot y$ of $x \in G_3^j$ and $y \in G_3^k$ is defined as 0 if $j = 0$ or $k = 0$ and otherwise as $(xy)_{k-j}$ *if* $j \leqslant k$ and $(xy)_{j-k}$ if $j \geqslant k$. For multivectors, it is just extended by bilinearity. For vectors, the key formula tells us that $\boldsymbol{x} \cdot \boldsymbol{y}$ is the scalar product, which is coherent with the notation that we have been using in this case as an alternative to $q(\boldsymbol{x}, \boldsymbol{y})$. For two bivectors, $\boldsymbol{x}\mathbf{i} \cdot \boldsymbol{y}\mathbf{i} = (\boldsymbol{x}\mathbf{i}\boldsymbol{y}\mathbf{i})_0 = (-\boldsymbol{x}\boldsymbol{y})_0 = -\boldsymbol{x} \cdot \boldsymbol{y}$, and in particular $\boldsymbol{x}\mathbf{i} \cdot \boldsymbol{x}\mathbf{i} = -\boldsymbol{x}^2$. For a vector and bivector, $\boldsymbol{x} \cdot \boldsymbol{y}\mathbf{i} = (\boldsymbol{x}\boldsymbol{y}\mathbf{i})_1 = (\boldsymbol{x} \wedge \boldsymbol{y})\mathbf{i}$, *while* $\boldsymbol{y}\mathbf{i} \cdot \boldsymbol{x} = (\boldsymbol{y}\mathbf{i}\boldsymbol{x})_1 = (\boldsymbol{y} \wedge \boldsymbol{x})\mathbf{i} = -\boldsymbol{x} \cdot \boldsymbol{y}\mathbf{i}$ (recall that $(\cdot)_k$ means the grade k component of $(\cdot)$).

The special case $\boldsymbol{w} \cdot (\boldsymbol{x} \wedge \boldsymbol{y})$ is important: its value is $\boldsymbol{w} \cdot (\boldsymbol{x} \wedge \boldsymbol{y}) = (\boldsymbol{w} \cdot \boldsymbol{x})\boldsymbol{y} - (\boldsymbol{w} \cdot \boldsymbol{y})\boldsymbol{x}$. For a proof, note that it is enough, due to the bilinearity in $\boldsymbol{x}$ and $\boldsymbol{y}$, to check the formula for $\boldsymbol{x} = \boldsymbol{e}_j$, $\boldsymbol{y} = \boldsymbol{e}_k$, $j \neq k$. In this case, $\boldsymbol{x} \wedge \boldsymbol{y} = \boldsymbol{e}_j\boldsymbol{e}_k$ and $\boldsymbol{w} \cdot (\boldsymbol{x} \wedge \boldsymbol{y}) = (\boldsymbol{w}\boldsymbol{e}_j\boldsymbol{e}_k)_1$. Writing $\boldsymbol{w} = w_1\boldsymbol{e}_1 + w_2\boldsymbol{e}_2 + w_3\boldsymbol{e}_3$, only the terms j and k in this sum count in the last expression, for the third term has grade 3, and

$$(\boldsymbol{w}\boldsymbol{e}_j\boldsymbol{e}_k)_1 = w_j\boldsymbol{e}_j\boldsymbol{e}_j\boldsymbol{e}_k + w_k\boldsymbol{e}_k\boldsymbol{e}_j\boldsymbol{e}_k = w_j\boldsymbol{e}_k - w_k\boldsymbol{e}_j = (\boldsymbol{w} \cdot \boldsymbol{e}_j)\boldsymbol{e}_k - (\boldsymbol{w} \cdot \boldsymbol{e}_k)\boldsymbol{e}_j.$$

1.3.15 (The cross product of vectors) The *cross product* $\boldsymbol{x} \times \boldsymbol{y} \in E_3$ of the vectors $\boldsymbol{x}$ and $\boldsymbol{y}$ is defined as the Hodge dual of $\boldsymbol{x} \wedge \boldsymbol{y}$:

$$\boldsymbol{x} \times \boldsymbol{y} = -(\boldsymbol{x} \wedge \boldsymbol{y})\mathbf{i} \ \text{ or } \ \boldsymbol{x} \wedge \boldsymbol{y} = (\boldsymbol{x} \times \boldsymbol{y})\mathbf{i}. \tag{1.7}$$

For example, with the cross product the second Olinde Rodrigues formula (see 1.3.11) becomes (in agreement with the original form)

$$\boldsymbol{u}'' \sin\tfrac{\theta''}{2} = \boldsymbol{u}\sin\tfrac{\theta}{2}\cos\tfrac{\theta'}{2} + \boldsymbol{u}'\cos\tfrac{\theta}{2}\sin\tfrac{\theta'}{2} + \boldsymbol{u}\times\boldsymbol{u}'\sin\tfrac{\theta}{2}\sin\tfrac{\theta'}{2}.$$

Note that whereas $\boldsymbol{x}\wedge\boldsymbol{y}$ does not depend on the orientation, $\boldsymbol{x}\times\boldsymbol{y}$ does, and in fact changes sign when the orientation is reversed. In old terminology, these kind of vectors were called *axial vectors,* while *polar vector* refers to the ordinary vectors (see Fig. 1.3). Actually, there is no mystery here, for $\boldsymbol{x}\times\boldsymbol{y}$ depends on $\boldsymbol{x}$, $\boldsymbol{y}$, and $\mathbf{i}$, so that a better notation for it would be $\boldsymbol{x}\times_{\mathbf{i}}\boldsymbol{y}$. If we do so for a moment, the "axiality" of the cross product is the relation $\boldsymbol{x}\times_{-\mathbf{i}}\boldsymbol{y} = -\boldsymbol{x}\times_{\mathbf{i}}\boldsymbol{y}$.

The properties of the cross product can be easily deduced with geometric algebra. Here is a brief indication of how it works.

(1) If j, k, l is a cyclic permutation of 1,2,3, then $\boldsymbol{e}_j\times\boldsymbol{e}_k = -\mathbf{i}(\boldsymbol{e}_j\wedge\boldsymbol{e}_k) = -\mathbf{i}(\mathbf{i}\boldsymbol{e}_l) = \boldsymbol{e}_l$. This shows that the cross product agrees with the usual linear algebra definition.
(2) By what has been established in 1.3.14, we have $\boldsymbol{x}\times\boldsymbol{y} = -\boldsymbol{x}\cdot\boldsymbol{y}\mathbf{i} = \boldsymbol{y}\mathbf{i}\cdot\boldsymbol{x}$. Since it is skew-symmetric, we also have $\boldsymbol{x}\times\boldsymbol{y} = -\boldsymbol{y}\times\boldsymbol{x} = -\boldsymbol{x}\mathbf{i}\cdot\boldsymbol{y} = \boldsymbol{y}\cdot\boldsymbol{x}\mathbf{i}$. In sum,

$$\boldsymbol{x}\times\boldsymbol{y} = -\boldsymbol{x}\mathbf{i}\cdot\boldsymbol{y} = -\boldsymbol{x}\cdot\boldsymbol{y}\mathbf{i} = \boldsymbol{y}\mathbf{i}\cdot\boldsymbol{x} = \boldsymbol{y}\cdot\boldsymbol{x}\mathbf{i}.$$

(3) The *mixed product formula,* $(\boldsymbol{x}\times\boldsymbol{y})\cdot\boldsymbol{z} = \det(\boldsymbol{x},\boldsymbol{y},\boldsymbol{z})$, can be deduced as follows. Since $\boldsymbol{x}\times\boldsymbol{y}$ is a vector, $2(\boldsymbol{x}\times\boldsymbol{y})\cdot\boldsymbol{z} = (\boldsymbol{x}\times\boldsymbol{y})\boldsymbol{z} + \boldsymbol{z}(\boldsymbol{x}\times\boldsymbol{y}) = -\mathbf{i}(\boldsymbol{x}\wedge\boldsymbol{y})\boldsymbol{z} - \mathbf{i}\boldsymbol{z}(\boldsymbol{x}\wedge\boldsymbol{y})$, which equals $-2\mathbf{i}\boldsymbol{x}\wedge\boldsymbol{y}\wedge\boldsymbol{z} = 2\det(\boldsymbol{x},\boldsymbol{y},\boldsymbol{z})$ (we have used that $b\boldsymbol{x}+\boldsymbol{x}b = 2b\wedge\boldsymbol{x}$ for any bivector b and vector $\boldsymbol{x}$).
(4) The *double cross product formula* $(\boldsymbol{x}\times\boldsymbol{y})\times\boldsymbol{z} = (\boldsymbol{x}\cdot\boldsymbol{z})\boldsymbol{y} - (\boldsymbol{y}\cdot\boldsymbol{z})\boldsymbol{x}$ is also easy:

$$(\boldsymbol{x}\times\boldsymbol{y})\times\boldsymbol{z} = -\mathbf{i}(\boldsymbol{x}\times\boldsymbol{y})\cdot\boldsymbol{z} = -(\boldsymbol{x}\wedge\boldsymbol{y})\cdot\boldsymbol{z} = \boldsymbol{z}\cdot(\boldsymbol{x}\wedge\boldsymbol{y}) = (\boldsymbol{z}\cdot\boldsymbol{x})\boldsymbol{y} - (\boldsymbol{z}\cdot\boldsymbol{y})\boldsymbol{z}.$$

(5) Geometrically, the cross-product $\boldsymbol{x}\times\boldsymbol{y}$ of two linearly independent vectors is determined by the following properties: (i) $\boldsymbol{x}\times\boldsymbol{y}$ is orthogonal to both $\boldsymbol{x}$ and $\boldsymbol{y}$; (ii) its length is $A(\boldsymbol{x},\boldsymbol{y})$, the area of the parallelogram defined by $\boldsymbol{x}$ and $\boldsymbol{y}$; and (iii) $\boldsymbol{x}, \boldsymbol{y}, \boldsymbol{x}\times\boldsymbol{y}$ is positively oriented basis. The property (i) follows from Hodge duality or by applying the mixed product formula. The property (iii) is also a consequence of that formula, because $\det(\boldsymbol{x},\boldsymbol{y},\boldsymbol{x}\times\boldsymbol{y}) = (\boldsymbol{x}\times\boldsymbol{y})\cdot(\boldsymbol{x}\times\boldsymbol{y}) > 0$. As for (ii), we may assume that $\boldsymbol{x}$ and $\boldsymbol{y}$ are linearly independent, and by 1.3.14 we have $(\boldsymbol{x}\times\boldsymbol{y})^2 = -(\boldsymbol{x}\wedge\boldsymbol{y})^2$. Now $\boldsymbol{x}\wedge\boldsymbol{y} = |\boldsymbol{x}||\boldsymbol{y}|\,\boldsymbol{a}\sin\alpha$, where $\boldsymbol{a}$ denotes a unit area in $\langle\boldsymbol{x},\boldsymbol{y}\rangle$ (hence $\boldsymbol{a}^2 = -1$) and $\alpha = \alpha(\boldsymbol{x},\boldsymbol{y})$. Consequently $(\boldsymbol{x}\wedge\boldsymbol{y})^2 = |\boldsymbol{x}|^2|\boldsymbol{y}|^2\sin^2\alpha = A(\boldsymbol{x},\boldsymbol{y})^2$. □

1.3.16 (When do two quaternions commute?) Since real scalars commute with any quaternion, the question is posed for two non-real quaternions, say $h = \lambda + \boldsymbol{u}\mathbf{i}\mu$ and $h' = \lambda' + \boldsymbol{u}'\mathbf{i}\mu'$, where $\boldsymbol{u}$ and $\boldsymbol{u}'$ are unit vectors and $\mu\mu' \neq 0$. Then we have, using the key formula, $hh' = \lambda\lambda' + \boldsymbol{u}'\mathbf{i}\lambda\mu' + \boldsymbol{u}\mathbf{i}\lambda'\mu - (\boldsymbol{u}\cdot\boldsymbol{u}')\mu\mu' - (\boldsymbol{u}\wedge\boldsymbol{u}')\mu\mu'$.

With a similar expression for $h'h$, we get that $h'h - hh' = 2\boldsymbol{u} \wedge \boldsymbol{u}'\mu\mu'$. Therefore $h'h = hh'$ if and only if $\boldsymbol{u} \wedge \boldsymbol{u}' = 0$, which only happens when $\boldsymbol{u}' = \pm\boldsymbol{u}$. □

1.4 GA of the Minkowski Space, $\mathcal{G}_{1,3}$

This section is devoted to a mathematical presentation of the geometric algebra $\mathcal{G}_{1,3}$ of the *Minkowski space* $E_{1,3}$, but its geometry and physical applications will be considered in Chap. 3. For historical reasons that will be explained later, $\mathcal{G}_{1,3}$ is called the *Dirac algebra* and we will denote it by $\mathcal{D}$. Let us remark that this particular case may be a good preparation for the study of the general case (arbitrary non-singular signatures) because it has many of the required ingredients for its treatment (see [97]).

The existence of such an algebra, with the assumptions specified below, can be proved by means of the *Dirac representation*, much as we proved the existence of $\mathcal{G}_2$ by means of $\mathbb{R}(2)$ and that of $\mathcal{G}_3$ by means of $\mathbb{C}(2)$ (see E.1.8, p. 32). For a detailed conceptual proof, see, for example, [97]. Our main concern, therefore, will be the analysis of the rich structure displayed by $\mathcal{D}$, including its deep relation to $\mathcal{G}_3$, and conclude, in particular, that $\mathcal{D}$ is unique up to a natural isomorphism.

We will denote by $E = E_{1,3}$ a vector space of dimension 4 endowed with a metric that will be denoted by η (instead of the symbol q used for the Euclidean spaces). The subindexes indicate that the *signature* of η is $(1, 3)$. This means that there exist an orthogonal basis e_0, e_1, e_2, e_3 such that

$$\eta(e_0) = 1, \ \eta(e_1) = \eta(e_2) = \eta(e_3) = -1. \tag{1.8}$$

Any such basis will be said to be *orthonormal*. Note that vectors are not written in bold italic characters, but just in italic lowercase letters. We reserve the bold italic for the 3D Euclidean space of relative vectors that will be introduced later.

A vector $a \in E$ will be said to be *positive* (*negative*) if $\eta(a) > 0$ ($\eta(a) < 0$). If $\eta(a) = 0$, then we say that a is *isotropic* or *null* (the 0 vector is null, of course, but there are plenty of non-zero vectors that are null, as, for example, $e_0 - e_k$, $k = 1, 2, 3$).

Let us remark that the signature is well defined: in any orthogonal basis of $E_{1,3}$, one of its vectors is positive and the remaining three are negative (see E.1.7, p. 32). As in the previous sections, the main assumption on $\mathcal{D}$ is that it is an algebra that extends E (we will use the same conventions as before about its product, called *geometric product*) and that it satisfies *Clifford's reduction rule* for any vector a: $a^2 = \eta(a)$. This rule is equivalent to *Clifford's relations*: $ab + ba = 2\eta(a, b)$ for all vectors a and b. The proof of the equivalence is the same as for the Euclidean case.

In addition, we require that $\mathbb{R} \cap E = \{0\}$ and that $\mathcal{D}$ is generated by E as an $\mathbb{R}$-algebra.

In what follows, e_0, e_1, e_2, e_3 denote an orthonormal basis of E. To refer to this basis we will use the symbol $\mathbf{e}$.

Clifford Units

For any sequence $J = j_1, \ldots, j_m \in N = \{0, 1, 2, 3\}$, set $e_J = e_{j_1} \cdots e_{j_m}$. We also define $\eta_J = \eta(e_{j_1}) \cdots \eta(e_{j_m}) = (-1)^{s(J)}$, where $s(J)$ is the number of indices l such that $\eta(e_{j_l}) = -1$.

Among the expressions e_J, we are especially interested in those for which J is a *multiindex*, which means that $j_1 < \cdots < j_m$, and in this case the e_J will be called *Clifford units*. Note that the set $\mathcal{J}$ of multiindices has cardinal $2^4 = 16$.

1.4.1 (Artin's formula [4])

(1) *If* $I, J \in \mathcal{J}$,

$$e_I e_J = (-1)^{t(I,J)}\, \eta_{I \cap J}\, e_{I \,\triangle\, J},$$

where $I \triangle J$ *is the* (ordered) *symmetric difference of* I *and* J *and* $t(I, J)$ *is the number of transpositions* (order inversions) *in the concatenated sequence* I, J.

(2) *In particular,* $e_J^2 = (-1)^{m/\!\!/2} \eta_J = (-1)^{s(J)+m/\!\!/2}$*, where* $m = |J|$ *and* $m/\!\!/2 = \lfloor m/2 \rfloor$ (the integer part of $m/2$).

Proof (1) The sign $(-1)^{t(I,J)}$ is the result of repeatedly applying the anticommutation rule for orthogonal vectors until reordering I, J in non-decreasing order. The sign $\eta_{I \cap J}$ is the result of applying the contraction rule to repeated vectors. What is left is clearly $e_{I \,\triangle\, J}$.

(2) We have $t(J, J) = \binom{m}{2}$ and $\eta_J = (-1)^{s(J)}$. Therefore $e_J^2 = (-1)^{s(J)+\binom{m}{2}}$. Now it is enough to observe that $\binom{m}{2}$ has the same parity as $m/\!\!/2$. □

As in the Euclidean spaces, the *volume element* or *pseudoscalar*

$$\mathbf{i} = e_0 e_1 e_2 e_3 = e_{0123} \tag{1.9}$$

will play an important role. For the moment, let us illustrate (2) in the proposition above: $\mathbf{i}^2 = -1$, as $s(0123) = 3$ and $4/\!\!/2 = 2$.

Another example is that if $e_{ijk}^2 = 1$, then necessarily $ijk = 123$. Indeed, since $3/\!\!/2 = 1$, $s(ijk)$ has to be odd and this can only happen if 0 is not in ijk.

To continue, notice that $\mathbf{i}$ *anticommutes with vectors*. This follows from the fact that it anticommutes with the elements of $\mathbf{e}$, an assertion that is straightforward to check. For example, $e_0\mathbf{i} = e_{123}$ while $\mathbf{i}e_0 = (-1)^3 e_0^2 e_{123} = -e_{123}$.

The following two statements amount to a specific treatment for the signature $(1, 3)$ of concepts, results, and methods that are valid in general. The issues involved (some quite subtle) are discussed in detail in [97].

1.4.2 (The Clifford units are distinct) (1) *For any* $I \in \mathcal{J}$, $I \neq \emptyset$, *we have* $e_I \neq \pm 1$. (2) *If* $I, J \in \mathcal{J}$ *and* $I \neq J$, *then* $e_I \neq \pm e_J$.

Proof (1) If $e_I = \pm 1$, $e_I^2 = 1$. This rules out the I such that $e_I^2 = -1$. The remaining I such that $|I| \leqslant 2$ are also ruled out, for $e_0 \neq \pm 1$, and if we had $e_j e_k = \pm 1$ ($j \neq k$), then we would get the contradiction $e_j = \pm e_k$. So the only remaining case to be ruled out is e_{123}. If it were $e_{123} = \pm 1$, multiplying by e_0 on the left we would get $\mathbf{i} = \pm e_0$, which cannot occur because $\mathbf{i}^2 = -1$ and $(\pm e_0)^2 = e_0^2 = 1$.

(2) The equality $e_I = \pm e_J$ implies that $\pm 1 = e_I^2 = \pm e_I e_J = \pm e_{I \,\triangle\, J}$, which by (1) is only possible if $I \triangle J = \emptyset$, that is, only if $I = J$. □

1.4.3 (Clifford basis) *The set* $B = \{e_J \mid J \in \mathcal{J}\}$ *of Clifford units is a linear basis of* $\mathcal{D}$. *Thus* $\dim \mathcal{D} = 16$.

Proof As in the Euclidean cases, we see that the geometric product of any number of vectors is a linear combination of B. Since E generates $\mathcal{D}$ as an $\mathbb{R}$-algebra, we have $\mathcal{D} = \langle B \rangle$. So it will suffice to prove that B is linearly independent.

Suppose that we have a linear relation $\sum_I \lambda_I e_I = 0$. We want to show that $\lambda_I = 0$ for all I. To that end, it will be enough to show that $\lambda_\emptyset = 0$. Indeed, 1.4.2 tells us that if we multiply the initial relation by an arbitrary e_I, then we obtain a similar relation whose $e_\emptyset$ coefficient is $\pm\lambda_I$.

So let us show that $\lambda_\emptyset = 0$. For each index k, the original relation clearly implies that $\sum_I \lambda_I e_k e_I e_k^{-1} = 0$. Since e_k commutes or anticommutes with e_I, we easily infer the relation $\sum_I \lambda_I e_I = 0$ in which the sum is extended to all the e_I *that commute with all the* e_k. Finally, note that e_I anticommutes with anyone of its factors when $|I|$ is even and positive, and that it anticommutes with any e_k such that $k \notin I$ when $|I|$ is odd. Since such k exist (any $k \in N - I \neq \emptyset$, as $|N| = 4$), we are just left with the relation $\lambda_\emptyset = 0$. □

Exterior Product and the Canonical Linear Grading of $\mathcal{D}$

For any $k \in N$, $k \neq 0$, consider the map $A : E^k \to \mathcal{D}$ such that

$$A(x_1, \ldots, x_k) = \tfrac{1}{k!}\textstyle\sum_p (-1)^{t(p)} x_{p_1} \cdots x_{p_k},$$

where the sum is extended to all permutations p of $\{1, \ldots, k\}$ and where $t(p)$ is the number of transpositions (order inversions) in p. This map is k-multilinear and skew-symmetric, and hence it induces a unique linear map $\mathsf{g} : \wedge^k E \to \mathcal{D}$ such that

$$\mathsf{g}(x_1 \wedge \cdots \wedge x_k) = A(x_1, \ldots, x_k). \tag{1.10}$$

Note that $\mathsf{g}(x_1 \wedge \cdots \wedge x_k) = x_1 \cdots x_k$ if $x_1, \ldots, x_k$ are pairwise orthogonal, for in that case the geometric product is skew-symmetric and $(-1)^{t(p)} x_{p_1} \cdots x_{p_k} = x_1 \cdots x_k$

for all permutations p. In particular, if we let $e_{\hat{J}} = e_{j_1} \wedge \cdots \wedge e_{j_k}$, then $\mathsf{g}(e_{\hat{J}}) = e_J$ for all J.

Since the k-vectors $e_{\hat{J}}$ for $J \in \mathcal{J}_k = \{J \in \mathcal{J} : |J| = k\}$ form a basis of $\wedge^k E$, we conclude that g is a canonical linear isomorphism $\wedge^k E \simeq \langle B_k \rangle$, where $B_k = \{e_J : J \in \mathcal{J}_k\}$. In particular we see that the spaces $\mathcal{D}^k = \langle B_k \rangle$ do not depend on the orthonormal basis $\mathbf{e}$ used to construct the Clifford basis and that we have a canonical decomposition $\mathcal{D} = \mathcal{D}^0 + \mathcal{D}^1 + \mathcal{D}^2 + \mathcal{D}^3 + \mathcal{D}^4$.

Now, again as in the Euclidean case, the linear isomorphism $\wedge E \simeq \mathcal{D}$ allows us to graft the structures of $\wedge E$ to $\mathcal{D}$. In particular, we can endow $\mathcal{D}$ with an exterior product (denoted with the same symbol $\wedge$, it is also called *outer product*), which enriches $\mathcal{D}$ with another algebra structure (associative and unital). For the computation of this exterior product, the most basic formula is that $e_{\hat{J}} = e_J$. In addition, the exterior product is *graded* and *skew-commutative*, which means that if $x \in \mathcal{D}^k$ and $y \in \mathcal{D}^l$, then $x \wedge y \in \mathcal{D}^{k+l}$ and $x \wedge y = (-1)^{kl} y \wedge x$.

The linear isomorphism $\wedge E \simeq \mathcal{D}$ also entitles us to apply the usual terminology concerning $\wedge E$. The elements of $\mathcal{D}$, for example, will be called *multivectors* and those of $\mathcal{D}^k$, k-vectors. We also say that k is the *grade* of the elements of $\mathcal{D}^k$. The non-zero k-vectors of the form $a_1 \wedge \cdots \wedge a_k$ are called *k-blades* (or also *decomposable* k-vectors). The 0-vectors are the *scalars*, as $\mathcal{D}^0 = \langle 1 \rangle = \mathbb{R}$. The 1-vectors are the usual *vectors*, for $\mathcal{D}^1 = \langle e_0, e_1, e_2, e_3 \rangle = E$. Instead of 2-vectors and 3-vectors, it is customary to say *bivectors* and *trivectors*. The 4-vectors form a 1-dimensional space, for $\mathcal{D}^4 = \langle \mathbf{i} \rangle$, and its elements are called *volume elements* or *pseudoscalars*.

Another important concept that can be transferred from $\wedge E$ to $\mathcal{D}$ is the natural extension of the metric η to $\wedge E$ and which we will still denote η. As seen in $\mathcal{D}$, this metric is determined by two conditions: The spaces $\mathcal{D}^k$ and $\mathcal{D}^l$ are orthogonal for $k \neq l$, and for two k-blades it is given by *Gram's formula*, which here it will be sufficient to state for $k = 2$:

$$\eta(a_1 \wedge a_2, a_1' \wedge a_2') = \begin{vmatrix} \eta(a_1, a_1') & \eta(a_1, a_2') \\ \eta(a_2, a_1') & \eta(a_2, a_2') \end{vmatrix}, \quad \eta(a_1 \wedge a_2) = \begin{vmatrix} \eta(a_1) & \eta(a_1, a_2) \\ \eta(a_2, a_1) & \eta(a_2) \end{vmatrix}. \tag{1.11}$$

In particular we have $\eta(e_I, e_J) = 0$ if $I \neq J$, and $\eta(e_I) = (-1)^{s(I)}$. Thus *the Clifford basis is orthonormal*. On the other hand, if we compare the value of e_J^2 with $\eta(e_J)$, we get the relation

$$e_J^2 = (-1)^{m/\!/2} \eta(e_J), \; m = |J|. \tag{1.12}$$

The Inner Product

The algebra $\mathcal{D}$ is endowed with another bilinear product, denoted by $x \cdot y$ and called the *inner product*. It is not associative, nor unital, but it turns out to be a fundamental ingredient. Due to the bilinearity, it is enough to define the inner product $e_I \cdot e_J$ of two Clifford units. The fact that it does not depend on the Clifford units used will be an immediate consequence of the definition in terms of the geometric product and the grading (see 1.4.4).

Let $l = |I|$ and $m = |J|$. The rules for the computation of $e_I \cdot e_J$ are as follows: If $l = 0$ or $m = 0$, then $e_I \cdot e_J = 0$; in other words, we have $1 \cdot e_J = e_I \cdot 1 = 0$ (this rule might seem a bit odd, but it is expedient in order to guarantee the validity without exceptions of important formulas that we will find later).

If $l, m \geqslant 1$, then $e_I \cdot e_J = e_I e_J$ if $I \subseteq J$ or $J \subseteq I$, and $= 0$ in all other cases. The explanation of this rule is that if we fix l and m, then the grade of $e_I e_J$ is $l + m - 2\nu$, where $\nu = |I \cap J|$, and so its minimum possible grade is when ν is maximum, which occurs precisely when $I \subseteq J$ (and then its grade is $m - l$) or $J \subseteq I$ (and then the grade is $l - m$). To sum up:

$$e_I \cdot e_J = \begin{cases} 0 & \text{if } l = 0 \text{ or } m = 0 \\ (e_I e_J)_{|l-m|} & \text{if } l, m \geqslant 1. \end{cases} \tag{1.13}$$

In particular we have $e_I \cdot e_I = e_I^2$ if $l \geqslant 1$.

Notice that the maximum possible grade of $e_I e_J$ is $l + m$, and that it is reached if and only if $\nu = 0$, which is to say if and only if $I \cap J = \emptyset$. If that is the case, $e_I e_J = e_I \wedge e_J$.

If we interchange the factors of the inner product (1.13) in the case $I \subseteq J$ ($J \subseteq I$), the number of sign changes is $(m-1)l = ml - l$ (respectively $(l-1)m = lm - m$).

All the considerations so far are systematized in the following statement:

1.4.4 (Grades of a product) *Let $x \in \mathcal{D}^l$, $y \in \mathcal{D}^m$. If $j \in \{0, 1, 2, 3, 4\}$ and $(xy)_j \neq 0$, then $j = |m - l| + 2\nu$ with $\nu \geqslant 0$ and $j \leqslant r + s$. Moreover, $(xy)_{l+m} = x \wedge y$ and for $l, m > 0$, $(xy)_{|l-m|} = x \cdot y$. Finally, $x \cdot y = (-1)^{lm+l} y \cdot x$ if $l \leqslant m$ and $= (-1)^{lm+m} y \cdot x$ if $m \leqslant l$.* □

1.4.5 (Remark) We have systematically used the metric η in order to avoid confusions with the inner product. A remarkable difference is that if $x \in \mathcal{D}^l$ and $y \in \mathcal{D}^m$, then $\eta(x, y) = 0$ when $l \neq m$, but in general $x \cdot y$ may be $\neq 0$ and also may be non-symmetric. For example, $e_1 \cdot e_0 e_1 e_2 = e_0 e_1 e_2 \cdot e_1 = e_0 e_2$, but $e_1 \cdot e_0 e_1 e_2 e_3 = e_0 e_2 e_3$ while $e_0 e_1 e_2 e_3 \cdot e_1 = -e_0 e_2 e_3$. In the case $l = m$, we have $x \cdot y = y \cdot x$ and we will see in 1.4.8 that this is equal to $(-1)^{m/\!/2} \eta(x, y)$. For example, $e_1 e_2 \cdot e_1 e_2 = e_1 e_2 e_1 e_2 = -1$ and $\eta(e_1 e_2) = \eta(e_1)\eta(e_2) = 1$.

1.4.6 (Key formulas) *If a is a vector and x a multivector, then*

$$ax = a \cdot x + a \wedge x \quad \textit{and} \quad xa = x \cdot a + x \wedge a.$$

Proof Because of the bilinearity in a and x, we may assume that $a = e_j$ and $x = e_K$. If $K = \emptyset$ the inner products vanish and both the geometric and the exterior products are equal to e_j. If $K \neq \emptyset$, there are two cases to consider: $j \in K$ and

$j \notin K$. If $j \in K$, $e_j \cdot e_K = e_j e_K$ and $e_K \cdot e_j = e_K e_j$, while $e_j \wedge e_K = e_K \wedge e_j = 0$. If $j \notin K$, $e_j \cdot e_K = e_K \cdot e_j = 0$, while $e_j e_K = e_j \wedge e_K$ and $e_K e_j = e_K \wedge e_j$. □

Involutions

Now let us transfer the parity and reverse involutions of $\wedge E$ to $\mathcal{D}$.

So we have parity and reverse involutions $\mathcal{D} \to \mathcal{D}$, $x \mapsto \hat{x}$ and $x \mapsto \tilde{x}$, which are given, for $x \in \mathcal{D}^k$, by $\hat{x} = (-1)^k x$ and $\tilde{x} = (-1)^{k/\!/2} x$, respectively.

1.4.7 (Properties of the involutions) *The parity involution is an* automorphism *of $\mathcal{D}$, in the sense that*

$$\widehat{xy} = \hat{x}\,\hat{y}, \quad \widehat{x \wedge y} = \hat{x} \wedge \hat{y}, \quad \widehat{x \cdot y} = \hat{x} \cdot \hat{y}.$$

The reverse involution is an antiautomorphism *of $\mathcal{D}$, in the sense that*

$$\widetilde{xy} = \tilde{y}\,\tilde{x}, \quad \widetilde{x \wedge y} = \tilde{y} \wedge \tilde{x}, \quad \widetilde{x \cdot y} = \tilde{y} \cdot \tilde{x}.$$

Proof In all cases, it is enough to check the identities for two elements of the Clifford basis, say $x = e_I \in \mathcal{D}^l$, $y = e_J \in \mathcal{D}^m$. For the parity involution, note that the degrees of $e_I e_J$, $e_I \wedge e_J$, and $e_I \cdot e_J$ are $l + m - 2\nu$ ($\nu = |I \cap J|$), $l + m$ and $|l - m|$, respectively, and that all are congruent to $l + m$ mod 2. For example, the right-hand side of $\widehat{xy} = \hat{x}\,\hat{y}$ is $(-1)^{l+m} xy$ and the left-hand side is $(-1)^{l+m-2\nu} xy$, so they are equal.

In the case of the reverse involution, the argument is similar, but we have to take into account that $\tilde{e}_I = (-1)^{l/\!/2} e_I = e_{\tilde{I}}$, where $\tilde{I}$ is the reverse of I (the order reversal $\tilde{I}$ of I amounts to $\binom{l}{2}$ sign changes, and we know that it is the same as $l/\!/2$ sign changes). In the case of the last equality, for example, and assuming that $l \leqslant m$, the right-hand side is $(-1)^{l/\!/2+m/\!/2} y \cdot x = (-1)^{l/\!/2+m/\!/2+lm+l} x \cdot y$, whereas the left-hand side is $(-1)^{(m-l)/\!/2} x \cdot y$, and they agree because a little arithmetical checking shows that $l/\!/2 + m/\!/2 + lm + l$ and $(m - l)/\!/2$ are congruent mod 2. □

1.4.8 (Alternative form of the metric) (1) *For any $x, y \in \mathcal{D}$, $\eta(x, y) = (x\tilde{y})_0 = (\tilde{x}y)_0$.* (2) *If X is a blade of grade k, then $\eta(X) = (-1)^{k/\!/2} X^2$. In particular, $X^2 \in \mathbb{R}$.*

Proof (1) Given that the three expressions are bilinear, it suffices to show that they hold for $x = e_I$ and $y = e_J$. Thus the checking is reduced to ascertain that $(e_I \tilde{e}_J)_0$ and $(\tilde{e}_I e_J)_0$ vanish if $J \neq I$, and that they are equal to $\eta(e_I)$ if $J = I$.

The first claim holds because the grades of $e_I \tilde{e}_J$ and $\tilde{e}_I e_J$ are not 0 if $J \neq I$. Concerning the second, it is clear that $e_I \tilde{e}_I = \tilde{e}_I e_I = (-1)^{s(I)}$ and we know that this is the value of $\eta(e_I)$.

(2) We have $\eta(X) = (X\tilde{X})_0$. But $X\tilde{X} \in \mathbb{R}$, because we can represent X as a product of orthogonal vectors, and so $\eta(X) = X\tilde{X} = (-1)^{k/\!/2} X^2$. □

Hodge Duality

Artin's formula 1.4.1 shows that if e_I has grade l, then $e_I\mathbf{i}$ has grade $4 - l$. So we have a linear map

$$\mathcal{D}^l \to \mathcal{D}^{4-l}, \; x \mapsto x^* = x\mathbf{i}.$$

This map, which is called *Hodge duality*, is a linear isomorphism, and its inverse is the map $y \mapsto -y\mathbf{i}$. Even more:

1.4.9 *The Hodge duality is an antiisometry.*

Proof $\eta(x^*) = \eta(x\mathbf{i}) = (x\mathbf{i}\,\widetilde{x\mathbf{i}})_0 = -(x\tilde{x})_0 = -\eta(x)$. □

Relative Space and the Pauli Algebra

Let $\mathcal{D}^+ = \mathcal{D}^0 + \mathcal{D}^2 + \mathcal{D}^4$. Since $\mathcal{D}^+ = \{x \in \mathcal{D} : \hat{x} = x\}$ and $x \mapsto \hat{x}$ is an automorphism of $\mathcal{D}$, $\mathcal{D}^+$ is a subalgebra of $\mathcal{D}$ that is called the *even subalgebra*. By what we have seen at the end of the previous subsection, we have

$$\mathcal{D}^2 = \langle \sigma_1, \sigma_2, \sigma_3, \sigma_1\mathbf{i}, \sigma_2\mathbf{i}, \sigma_3\mathbf{i} \rangle = \mathcal{E} + \mathcal{E}\mathbf{i},$$

where $\mathcal{E}$ is the space $\langle \sigma_1, \sigma_2, \sigma_3 \rangle$. Since $\sigma_k^2 = -\eta(\sigma_k) = 1$, we see that $\mathcal{E}$ is a 3-dimensional Euclidean space with the metric $q = -\eta|_{\mathcal{E}}$ and that $\{\sigma_1, \sigma_2, \sigma_3\}$ is a q-orthonormal basis. We will say that $\mathcal{E}$ is the *relative space* associated with e_0. The geometrical and physical significance of this space will be considered in Chap. 3.

1.4.10 *$\mathcal{D}^+$ is the geometric algebra of $(\mathcal{E}, q)$ and its pseudoscalar is* $\mathbf{i}$.

Proof $\mathcal{D}^+$ is an associative unital algebra, it contains $\mathcal{E}$ as a subspace (Fig. 1.4), and $\mathbb{R} \cap \mathcal{E} = \{0\}$. It is also immediate that $\mathcal{D}^+$ is generated by $\mathcal{E}$ as an $\mathbb{R}$-algebra, for $\sigma_j\mathbf{i} = \sigma_k\sigma_l$ (where jkl is a cyclic permutation of 123) and $\mathbf{i} = \sigma_1\sigma_2\sigma_3$. Finally, the fact that $\{\sigma_1, \sigma_2, \sigma_3\}$ is a q-orthonormal basis of $\mathcal{E}$, and that $\sigma_k^2 = 1 = q(\sigma_k)$, shows

Grade	Names	Bases
0	*Scalars*	1
1	*Vectors*	$e_0, \overline{e}_1, \overline{e}_2, \overline{e}_3$
2	*Bivectors*	$\overline{\sigma}_1, \overline{\sigma}_2, \overline{\sigma}_3, \sigma_1^*, \sigma_2^*, \sigma_3^*$
3	*Pseudovectors*	$\overline{e}_0^*, e_1^*, e_2^*, e_3^*$
4	*Pseudoescalars*	$\overline{1}^*$

Fig. 1.4 Synopsis of the Clifford basis (up to a few signs) with a row for each grade. For grade 2, the basis is split in two parts: The bivectors $\sigma_k = e_k e_0$ and their Hodge duals σ_k^* ($k = 1, 2, 3$). The significance of the order $e_k e_0$ (instead of $e_0 e_k$) will be seen in the next subsection. For a basis element x, we write $\bar{x}$ to denote that $\eta(x) = -1$

that $\mathcal{D}^+$ satisfies the contraction rule with respect to q: $\boldsymbol{x}^2 = q(\boldsymbol{x})$ for any $\boldsymbol{x} \in \mathcal{E}$. The relation $\sigma_1\sigma_2\sigma_3 = \mathbf{i}$ shows that the pseudoscalar of $\mathcal{E}$ coincides with $\mathbf{i}$. □

We set $\mathcal{P} = \mathcal{D}^+$ and will say that it is the *Pauli algebra*. The linear grading of this algebra is as follows: $\mathcal{P}^0 = \mathbb{R}$, $\mathcal{P}^1 = \mathcal{E}$, $\mathcal{P}^2 = \mathcal{E}\mathbf{i}$, $\mathcal{P}^3 = \langle\mathbf{i}\rangle = \mathcal{D}^4$. As established in the preceding chapter, the even algebra of the Pauli algebra, $\mathcal{P}^+ = \mathcal{P}^0 + \mathcal{P}^2$, is the field of (geometric) quaternions.

1.4.11 (Remark) The geometric product of $\mathcal{P}$ is the restriction of the geometric product of $\mathcal{D}$. Moreover, $\mathcal{P}$ is clearly closed for the exterior and inner products of $\mathcal{D}$ (cf. 1.4.7). But the restrictions of the latter two products to $\mathcal{P}$ do not agree with the exterior and inner products of $\mathcal{P}$. To distinguish between the two exterior and inner products, we make the convention of writing $\sigma_1, \sigma_2, \sigma_3$ when we consider these expressions as bivectors of $\mathcal{D}$, and $\boldsymbol{\sigma}_1, \boldsymbol{\sigma}_2, \boldsymbol{\sigma}_3$ when they are considered as vectors of $\mathcal{E}$. In this way, an expression such as $\boldsymbol{\sigma}_1 \wedge \boldsymbol{\sigma}_2$ indicates that the exterior product is to be taken in $\mathcal{P}$, and the result is the bivector $\boldsymbol{\sigma}_1\boldsymbol{\sigma}_2 = \boldsymbol{\sigma}_3\boldsymbol{i} \in \mathcal{P}^2$ (we let $\boldsymbol{i}$ denote the pseudoscalar $\mathbf{i}$ of $\mathcal{D}$ when considered as the pseudoscalar of $\mathcal{P}$). Note, however, that $\sigma_1 \wedge \sigma_2 = e_1 \wedge e_0 \wedge e_2 \wedge e_0 = 0$ in $\mathcal{D}$. Analogously, $\boldsymbol{\sigma}_2 = \boldsymbol{\sigma}_1 \cdot \boldsymbol{\sigma}_1\boldsymbol{\sigma}_2$, *but* $\sigma_1 \cdot \sigma_1\sigma_2 = -e_1e_0 \cdot e_1e_2 = 0$. □

1.4.12 (Complex Structure of $\mathcal{D}$) $\mathbf{C} = \langle 1, \mathbf{i}\rangle = \mathcal{D}^0 + \mathcal{D}^4 = \mathcal{P}^0 + \mathcal{P}^3$ is a subalgebra of $\mathcal{P}$ and $\mathcal{D}$ (recall that $\boldsymbol{i} = \mathbf{i}$) which is isomorphic to the complex field $\mathbb{C}$. Its elements have the form $\alpha + \beta\mathbf{i}$ ($\alpha, \beta \in \mathbb{R}$) and we call them *complex scalars.*

The space $\mathcal{D}^1 + \mathcal{D}^3 = \mathcal{D}^1 + \mathcal{D}^1\mathbf{i}$ is closed for the multiplication by $\mathbf{i}$ and we will say that it is the space of *complex vectors.* Its elements have the form $a + b\mathbf{i}$, $a, b \in \mathcal{D}^1$.

The elements of the space $\mathcal{D}^2 = \mathcal{E} + \mathcal{E}\mathbf{i}$, which is also closed for the multiplication by complex scalars, have the form $\boldsymbol{x} + \boldsymbol{y}\mathbf{i}$, $\boldsymbol{x}, \boldsymbol{y} \in \mathcal{E}$.

To sum up, all multivectors of $\mathcal{D}$ can be represented in a unique way in the form $(\alpha + \beta\mathbf{i}) + (a + b\mathbf{i}) + (\boldsymbol{x} + \boldsymbol{y}\mathbf{i})$, for $\alpha, \beta \in \mathbb{R}$, $a, b \in \mathcal{D}^1$, $\boldsymbol{x}, \boldsymbol{y} \in \mathcal{E}$. □

1.5 Exercises

E.1.1 ($\sqrt{2}$ is irrational) Let a be a positive real number. Let

$$X = \{x \in \mathbb{R} \mid x > 0,\ x^n \leqslant a\}.$$

Then X is bounded above and if r is its least upper bound, then $r^n = a$ (or $r = \sqrt[n]{a}$).

$\sqrt{2} \notin \mathbb{Q}$: If $\sqrt{2} = a/b$, $a, b \in \mathbb{N}$, we could assume that a and b are relatively prime; then, $2 = a^2/b^2$, or $a^2 = 2b^2$, and hence $a = 2a'$, $2a'^2 = b^2$; therefore, we arrive at the contradiction that a and b are both even.

E.1.2 The exponential function e^x is defined by the formula

$$e^x = 1 + \frac{x}{1!} + \frac{x^2}{2!} + \frac{x^3}{3!} + \cdots = \sum_{k \geqslant 0} \frac{x^k}{k!}.$$

In the case $x = i\alpha$, $(i\alpha)^k = i^k \alpha^k$, and since $i^2 = -1$, $i^3 = -i$, $i^4 = 1$, $i^5 = i, \ldots$, we get *Euler's formula*:

$$e^{i\alpha} = 1 - \frac{\alpha^2}{2!} + \frac{\alpha^4}{4!} - \cdots + i\left(\frac{\alpha}{1!} - \frac{\alpha^3}{3!} + \frac{\alpha^5}{5!} - \cdots\right) = \cos\alpha + i\sin\alpha.$$

In particular we have $e^{i\pi/2} = i$, $e^{i\pi} = -1$, and $e^{2i\pi} = 1$. For a given $n \in \mathbb{N}$, the complex numbers $\xi_k = e^{2\pi i k/n}$ satisfy $(\xi_k)^n = e^{2\pi i k} = 1$, and since for $k = 0, \ldots, n-1$ they are distinct, it follows that they are the only solutions of the equation $\xi^n = 1$.

E.1.3 To show that we can find an orthogonal basis for the metric q we can proceed as follows (we will not use that q is non-degenerate). If $q(\boldsymbol{x}) = 0$ for all $\boldsymbol{x}$, then the relation $2q(\boldsymbol{x}, \boldsymbol{y}) = q(\boldsymbol{x}+\boldsymbol{y}) - q(\boldsymbol{x}) - q(\boldsymbol{y})$ shows that q is identically 0, and in this case all bases are orthogonal. So we may assume that there is a vector $\boldsymbol{e}_1$ such that $q(\boldsymbol{e}_1) \neq 0$. Then the kernel of the linear map $E \to \mathbb{R}$ given by $\boldsymbol{x} \mapsto q(\boldsymbol{e}_1, \boldsymbol{x})$ has dimension $n-1$ and by induction it has an orthogonal basis $\boldsymbol{e}_2, \ldots, \boldsymbol{e}_n$. Since its vectors are orthogonal to $\boldsymbol{e}_1$ (by construction), it follows that $\boldsymbol{e}_1, \boldsymbol{e}_2, \ldots, \boldsymbol{e}_n$ is an orthogonal basis of E.

E.1.4 (Cauchy-Schwarz inequality) Given the non-zero vectors $\boldsymbol{v}, \boldsymbol{v}' \in E_n$, for any $\lambda \in \mathbb{R}$ we have $q(\lambda\boldsymbol{v} - \boldsymbol{v}') \geqslant 0$, and $q(\lambda\boldsymbol{v} - \boldsymbol{v}') = 0$ can only happen if $\boldsymbol{v}' = \lambda\boldsymbol{v}$ for some λ. In this case, $q(\boldsymbol{v}, \boldsymbol{v}') = \lambda q(\boldsymbol{v}) = \lambda|\boldsymbol{v}|^2 = \pm|\boldsymbol{v}||\boldsymbol{v}'|$, where the sign $\pm$ corresponds to the cases $\lambda > 0$ and $\lambda < 0$.

If $\boldsymbol{v}$ and $\boldsymbol{v}'$ are linearly independent, then $q(\lambda\boldsymbol{v} - \boldsymbol{v}') = q(\lambda\boldsymbol{v} - \boldsymbol{v}', \lambda\boldsymbol{v} - \boldsymbol{v}') > 0$ for all λ. Using that q is bilinear, we conclude that $q(\boldsymbol{v})\lambda^2 - 2\lambda q(\boldsymbol{v}, \boldsymbol{v}') + q(\boldsymbol{v}') > 0$ for any λ, which can occur only if $q(\boldsymbol{v}, \boldsymbol{v}')^2 - q(\boldsymbol{v})q(\boldsymbol{v}') < 0$, or $q(\boldsymbol{v}, \boldsymbol{v}')^2 < |\boldsymbol{v}|^2|\boldsymbol{v}'|^2$. Finally this is equivalent to say that $-|\boldsymbol{v}||\boldsymbol{v}'| < q(\boldsymbol{v}, \boldsymbol{v}') < |\boldsymbol{v}||\boldsymbol{v}'|$. Therefore

$$-|\boldsymbol{v}||\boldsymbol{v}'| \leqslant q(\boldsymbol{v}, \boldsymbol{v}') \leqslant |\boldsymbol{v}||\boldsymbol{v}'|$$

in all cases, with equality on the right (left) precisely when $\boldsymbol{v}' = \lambda\boldsymbol{v}$ with $\lambda > 0$ $(\lambda < 0)$.

E.1.5 With the notations $z_\varphi = e^{i\varphi} \in \mathrm{P}^+$ and $\boldsymbol{u}_\alpha = \boldsymbol{e}_1 z_\alpha \in \mathrm{P}^-$, we can translate into an element-wise form the assertions in the statement and proof of 1.2.9. In particular we have the following relations: (1) $z_\varphi z_{\varphi'} = z_{\varphi+\varphi'}$, (2) $\boldsymbol{u}_\alpha z_\varphi = \boldsymbol{u}_{\alpha+\varphi}$ and $z_\varphi \boldsymbol{u}_\alpha = \boldsymbol{u}_{\alpha-\varphi}$, (3) $\boldsymbol{u}_\alpha \boldsymbol{u}_\beta = z_{\beta-\alpha}$.

E.1.6 (Existence of $\mathcal{G}_2$) Pick an orthonormal basis $\boldsymbol{e}_1, \boldsymbol{e}_2$ of E_2 and consider (using the notations of 1.1.3) the isomorphism $m : E_2 = \langle \boldsymbol{e}_1, \boldsymbol{e}_2 \rangle \simeq \langle \mathfrak{e}_1, \mathfrak{e}_2 \rangle \subset \mathbb{R}(2)$ determined by $\boldsymbol{e}_j \mapsto \mathfrak{e}_j$ $(j = 1, 2)$. If $\boldsymbol{v} = \lambda_1 \boldsymbol{e}_1 + \lambda_2 \boldsymbol{e}_2$, then

$$m(\boldsymbol{v})^2 = (\lambda_1 \mathfrak{e}_1 + \lambda_2 \mathfrak{e}_2)^2 = \lambda_1^2 + \lambda_2^2 = q(\boldsymbol{v}).$$

This and the fact that $\{I_2, \mathfrak{e}_1, \mathfrak{e}_2, \mathfrak{e}_1\mathfrak{e}_2\}$ is a linear basis of $\mathbb{R}(2)$ show that $\mathcal{G}_2$ exists.

E.1.7 (Sylvester's law of inertia) Given an orthogonal basis of $E_{1,3}$, let F be the vector subspace spanned by its positive vectors. Since all non-zero vectors of F are positive, and all non-zero vectors of $\langle \boldsymbol{e}_1, \boldsymbol{e}_2, \boldsymbol{e}_3 \rangle$ are negative, $F \cap \langle \boldsymbol{e}_1, \boldsymbol{e}_2, \boldsymbol{e}_3 \rangle = \{0\}$. This implies, as $E_{1,3}$ has dimension 4, that $\dim(F) \leqslant 1$. But it cannot be $F = \{0\}$, as otherwise all non-zero vectors would be negative. So $\dim(F) = 1$.

This argument can be easily adapted to prove the general *Sylvester's law of inertia*: if a metric η of a vector space has an orthogonal basis with r positive vectors, s negative vectors, and t null vectors, then any other orthogonal basis has r positive vectors, s negative vectors, and t null vectors.

E.1.8 (The Dirac representation) In 1928, Dirac introduced the famous matrices $\Gamma_\mu \in \mathbb{C}(4)$, [27], namely:

$$\Gamma_0 = \begin{pmatrix} \sigma_0 & 0 \\ 0 & -\sigma_0 \end{pmatrix}, \quad \Gamma_k = \begin{pmatrix} 0 & -\sigma_k \\ \sigma_k & 0 \end{pmatrix}.$$

The matrices Γ_μ satisfy the Clifford relations for the signature $\eta = (+,\ ,-,-)$:

$$\Gamma_\mu \Gamma_\nu + \Gamma_\nu \Gamma_\mu = 2\eta_{\mu\nu},$$

and so we have a representation $\mathcal{D} \to \mathbb{C}(4)$ such that $e_\mu \mapsto \Gamma_\mu$. This representation can be used to prove that the Dirac algebra $\mathcal{D}$ exists by adapting the argument used for the existence of $\mathcal{G}_3$. Note, however, that in this case the image of $\mathcal{D}$ is a 16-dimensional subalgebra of the 32-dimensional algebra $\mathbb{C}(4)$.

Chapter 2
Conformal Geometric Algebra

This chapter is devoted to a presentation of conformal geometric algebra (CGA) targeted to the sort of applications dealt with in Chaps. 4 (robotics) and 5 (molecular geometry). This means that the ground space will be the Euclidean space E_3 and that the algebra we will be working with is designed so that it can encode *all conformal transformations of* E_3 *in spinorial form.* Except for noting that conformal means angle-preserving, we can defer the necessary precisions to the most convenient moments in our exposition. Here are some references: Foundational paper: [64]; a nice expository memoir: [55]; computationally oriented: [49]; vision and graphics oriented: [54]; oriented conformal geometry: [12]; treatises: [20, 30, 63, 79].

2.1 Ground Notions

The *conformal closure* of the Euclidean space $E = E_3$ is defined as the space

$$\bar{E} = E \perp E_{1,1} \simeq E_{4,1},$$

where $E_{1,1}$ is a *hyperbolic plane*, that is, a two-dimensional vector space with a metric of signature $(1, 1)$. If we let q denote the metric of $\bar{E}$ (the same symbol used for the metric of E), then we will say that the geometric algebra $C = \mathcal{G}_{4,1}$ of $\bar{E}$ is the *conformal geometric algebra* (CGA) of E.

The algebra C is constructed from $\bar{E}$ in much the same way as the Dirac algebra $\mathcal{D}$ was constructed from $E_{1,3}$. It is an algebra that contains $\bar{E}$ as a linear subspace satisfying $\mathbb{R} \cap \bar{E} = \{0\}$, that is generated by $\bar{E}$ as an $\mathbb{R}$-algebra, and which satisfies *Clifford's reduction rule*,

$$a^2 = q(a) \text{ for all } a \in \bar{E}, \tag{2.1}$$

C. Lavor et al., *A Geometric Algebra Invitation to Space-Time Physics, Robotics and Molecular Geometry*, SpringerBriefs in Mathematics,
https://doi.org/10.1007/978-3-319-90665-2_2

or, equivalently, *Clifford's relations*,

$$ab + ba = 2q(a, b) \text{ for all } a, b \in \bar{E}. \tag{2.2}$$

Now C can be endowed with a canonical linear grading

$$C = C^0 + C^1 + C^2 + C^3 + C^4 + C^5 \tag{2.3}$$

such that $\dim C^k = \binom{5}{k}$, and hence $\dim C = 32$, with $C^0 = \mathbb{R}$ and $C^1 = \bar{E}$. This may be constructed with the Clifford units associated with an orthonormal basis of $\bar{E}$ formed with an orthonormal basis e_1, e_2, e_3 of E and an orthonormal basis $e, \bar{e}$ of $E_{1,1}$ (so $e \cdot \bar{e} = 0$, $e^2 = 1$ and $\bar{e}^2 = -1$). This works because the statement corresponding to 1.4.2 (1) is still true and then a proof of the linear independence can be written by adapting the proof of 1.4.2 (2).

This enables us to use the terms *multivectors* (in general they will be denoted by symbols such as x, y, etc.) and *k-vectors* (the grade k component of $x \in C$ will be denoted $x_k \in C^k$; general blades will be denoted with symbols such as X, Y, etc.). In C we also have *outer* (or *exterior*) and *inner* products ($x \wedge y$ and $x \cdot y$, respectively), and also parity and reversal involutions ($\hat{x}$ and $\tilde{x}$, respectively).

All these ingredients obey rules that are analogous to the corresponding rules for $\mathcal{D}$ and which can be proved by similar methods. So, for example, we have a determination of the exterior and inner products analogous to 1.4.4 and the remark 1.4.5 (in particular $q(a, b) = a \cdot b$ for all $a, b \in E$). We also have key formulas 1.4.6, and the alternative form of the metric

$$q(x, y) = (x\tilde{y})_0 = (\tilde{x}y)_0, \tag{2.4}$$

for all $x, y \in C$ (cf. 1.4.8). Rather than delving into more details here, it will be sufficient to indicate them, when required, along the way.

Instead of the basis $\{e, \bar{e}\}$ of $E_{1,1}$, it is convenient to use the basis $\{e_0, e_\infty\}$ defined as follows:

$$e_0 = \tfrac{1}{2}(e + \bar{e}), \; e_\infty = \bar{e} - e. \tag{2.5}$$

These vectors satisfy the relations

$$e_0^2 = e_\infty^2 = 0, \qquad e_0 \cdot e_\infty = -1. \tag{2.6}$$

Note also that

$$e_0 + \tfrac{1}{2}e_\infty = \bar{e}, \qquad e_0 - \tfrac{1}{2}e_\infty = e. \tag{2.7}$$

As in Chap. 1, the vectors in E will be represented with bold italic symbols with the exception of the vectors e_1, e_2, e_3 of an orthonormal basis. Then $\mathbf{e} = \{e_1, e_2, e_3, e_0, e_\infty\}$ is a basis of $\bar{E}$, but it is not orthonormal.

Fig. 2.1 Basis blades of C. The square of the eight blades $1 = e_\emptyset$, e_k, $e_{0\infty}$ and $e_{k0\infty}$ ($k = 1, 2, 3$) is 1, the square of the eight blades e_{jk}, e_{123}, $e_{jk0\infty}$ and $e_{1230\infty}$ ($0 \leqslant j < k \leqslant 3$) is -1, and the square of the remaining 16 blades is 0

r	Basis blades of grade r
0	1
1	$e_1, e_2, e_3, e_0, e_\infty$
2	$e_{12}, e_{13}, e_{23}, e_{10}, e_{20}, e_{30}, e_{1\infty}, e_{2\infty}, e_{3\infty}, e_{0\infty}$
3	$e_{123}, e_{120}, e_{130}, e_{230}, e_{12\infty}, e_{13\infty}, e_{23\infty}, e_{10\infty}, e_{20\infty}, e_{30\infty}$
4	$e_{120\infty}, e_{130\infty}, e_{230\infty}, e_{1230}, e_{123\infty}$
5	$I = e_{1230\infty}$ (pseudoscalar)

2.1.1 (Matrix of q) The matrix of q with respect to $\mathbf{e}$ is:

$\cdot$	e_1	e_2	e_3	e_0	e_∞
e_1	1	0	0	0	0
e_2	0	1	0	0	0
e_3	0	0	1	0	0
e_0	0	0	0	0	-1
e_∞	0	0	0	-1	0

In other words, $(\boldsymbol{v}+\lambda_0 e_0+\lambda_\infty e_\infty)\cdot(\boldsymbol{v}'+\lambda_0' e_0+\lambda_\infty' e_\infty) = \boldsymbol{v}\cdot\boldsymbol{v}' - (\lambda_0\lambda_\infty' + \lambda_\infty\lambda_0')$.

For any $j_1, \cdots, j_k \in N = \{1, 2, 3, 0, \infty\}$, we set $e_{j_1\cdots j_k} = e_{j_1} \wedge \cdots \wedge e_{j_k}$. If the indices $j_1, \ldots, j_k$ are distinct, then $e_{j_1\cdots j_k}$ is a blade, and if we further impose that $j_1 < \cdots < j_k$ (in the order declared for N), then we get a basis of C^k, which we will simply call *blade basis*. In Fig. 2.1 we can see this basis displayed with a row for each grade.

2.1.2 (Examples)

(1) The 2-blade $Z = e_{0\infty}$ is equal to $e\bar{e}$ (the pseudoscalar of $E_{1,1}$) :

$$e_{0\infty} = e_0 \wedge e_\infty = \tfrac{1}{2}(e+\bar{e}) \wedge (\bar{e}-e)\wedge = e \wedge \bar{e} = e\bar{e},$$

where in the last step we use that e and $\bar{e}$ are orthogonal.

(2) We have $Z^2 = (e\bar{e})^2 = -e^2\bar{e}^2 = 1$, $\tilde{Z} = -Z$, $Ze = -eZ = -\bar{e}$, $Z\bar{e} = -\bar{e}Z = -e$.

(3) Using the alternative form of the metric [see 1.4.8 (2)], we can easily find the squares of the basis blades. If the blade contains the indexes 0∞, its square is 1 for grades 2 and 3, and -1 for grades 4 and 5. Otherwise, if it contains 0 or ∞, its square is 0, and if it only contains indexes from $\{1, 2, 3\}$, then we are in the case of $\mathcal{G}_3$ and the square is 1 for grades 0 and 1, and -1 for grades 2 and 3. See Fig. 2.1 for a different way of listing these values.

2.1.3 (Remarks)

(1) All pairs of vectors in the basis $e_1, e_2, e_3, e_0, e_\infty$ are orthogonal, except for e_0 and e_∞. Thus, only in this case $Z = e_{0\infty} = e_0 \wedge e_\infty$ differs from the geometric product $e_0 e_\infty$:

$$e_0 e_\infty = e_0 \cdot e_\infty + e_0 \wedge e_\infty = -1 + e_{0\infty} = -1 + Z.$$

(2) $e_\infty e_0 = \widetilde{e_0 e_\infty} = -1 + \tilde{Z} = -1 - Z$.
(3) $e_0 Z = -Z e_0 = e_0,\ \ e_\infty Z = -Z e_\infty = -e_\infty$.
(4) $e_0 e_\infty e_0 = -2e_0,\ \ e_\infty e_0 e_\infty = -2e_\infty$.

The Hestenes' Embedding

The vectors in $\bar{E}$ will be called *conformal vectors*, or just *vectors* if the context allows it. Any conformal vector has the form

$$x = \boldsymbol{v} + \lambda e_0 + \mu e_\infty, \qquad \lambda, \mu \in \mathbb{R}. \tag{2.8}$$

The set $Q = \{x \in \bar{E} : x^2 = 0\}$ of null vectors is a cone with vertex at 0, for if $x \in Q$, then $\lambda x \in Q$ for all $\lambda \in \mathbb{R}$. Note that $e_0, e_\infty \in Q$. We will say that Q is the *null cone* of $\bar{E}$. We will also write $Q' = Q - \langle e_\infty \rangle$.

2.1.4 (Normalized form of a null vector) *If $x \in Q'$, then there exist a non-zero scalar λ and an $\boldsymbol{x} \in E$, uniquely determined by x, such that $x = \lambda(\boldsymbol{x} + e_0 + \frac{1}{2}\boldsymbol{x}^2 e_\infty)$.*

Proof With the notations of relation (2.8), we have $x^2 = \boldsymbol{v}^2 - 2\lambda\mu$. Therefore, $x \in Q$ if and only if $\boldsymbol{v}^2 = 2\lambda\mu$. We can have $x \in Q'$ only if $\lambda \neq 0$ ($\lambda = 0$ implies $\boldsymbol{v} = 0$ and so we would have the contradiction $x = \mu e_\infty \notin Q'$). Thus we can write $x = \lambda(\boldsymbol{x} + e_0 + \frac{\mu}{\lambda} e_\infty)$, where $\boldsymbol{x} = \boldsymbol{v}/\lambda$. Finally $\frac{\mu}{\lambda} e_\infty = \frac{\boldsymbol{v}^2/2\lambda}{\lambda} e_\infty = \frac{1}{2}\boldsymbol{x}^2 e_\infty$. □

This tells us that the section $\mathcal{H}$ of Q by the hyperplane of equation $\lambda = 1$ (or $x \cdot e_\infty = -1$) is the locus of the vectors $x = \boldsymbol{x} + e_0 + \frac{1}{2}\boldsymbol{x}^2 e_\infty$ for $\boldsymbol{x} \in E$. Therefore, $\mathcal{H}$ is the paraboloid in the hyperplane $\lambda = 1$ of equation $\mu = \frac{1}{2}\boldsymbol{x}^2$ (often called the *horosphere*) and the map $H : E \to \mathcal{H}, \boldsymbol{x} \mapsto x$, defined by

$$x = \boldsymbol{x} + e_0 + \tfrac{1}{2}\boldsymbol{x}^2 e_\infty \tag{2.9}$$

is bijective (the inverse map $x \mapsto \boldsymbol{x}$ is given by the restriction to $\mathcal{H}$ of the orthogonal projection $\pi : \bar{E} \to E$ of $\bar{E}$ onto E). H is called the *Hestenes' map* (see Fig. 2.2).

2.1.5 (Remark on e_0 and e_∞) Since $e_0 = H(\boldsymbol{0})$, e_0 is the conformal vector of the origin of E. On the other hand, we have $\lim_{|\boldsymbol{x}|\to\infty} 2x/\boldsymbol{x}^2 = e_\infty$, which suggests that e_∞ is the conformal vector of the "point at infinity" of E. For a more precise geometric interpretation of this, see next remark. □

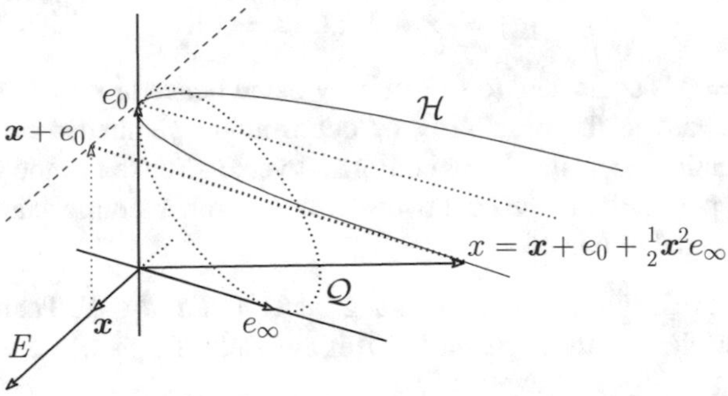

Fig. 2.2 The *Hestenes map*. The section $\mathcal{H}$ of the null cone Q by the hyperplane of $\bar{E}$ through e_0 which is parallel to $E + \langle e_\infty \rangle$ (this hyperplane is given by the equation $x \cdot e_\infty = -1$) is a paraboloid in that hyperplane (the *horosphere*). The Hestenes' vector x corresponding to $\boldsymbol{x} \in E$ belongs to $\mathcal{H}$ and the map $E \to \mathcal{H}$, $\boldsymbol{x} \mapsto x = H(\boldsymbol{x})$ is bijective. In this picture the section of Q by the hyperplane through e_0 that is parallel to $E + \langle e_0 - e_\infty \rangle$ is represented by the dotted ellipse

2.1.6 (Remark on conformal compactifications) Consider the projective space $\mathbf{P}\bar{E}$ of $\bar{E}$. Given a non-zero vector $x \in \bar{E}$, let $|x\rangle \in \mathbf{P}\bar{E}$ be the point defined by x. Recall that $|x\rangle$ is the line $\langle x \rangle \subset \bar{E}$ regarded as a point of $\mathbf{P}\bar{E}$, which gives the basic rule that $|x\rangle = |x'\rangle$ if and only if $x' = \lambda x$ for some $\lambda \in \mathbb{R}$ (necessarily non-zero because $x' \neq 0$), a relation that we denote $x \sim x'$.

Now sitting in $\mathbf{P}\bar{E}$ we have $\mathbf{P}Q$, which is the set of points $|x\rangle$ for $x \in Q - \{0\}$ (a quadric hypersurface in projective geometry terminology). The main use of this is that $\mathbf{P}\mathcal{H} = \mathbf{P}Q - \{|e_\infty\rangle\}$ and that the map $E \to \mathbf{P}\mathcal{H}$, $\boldsymbol{x} \mapsto |x\rangle$, is bijective. Thus E is embedded in the compact manifold $\mathbf{P}Q$ and fills it except for the point $|e_\infty\rangle$. This is why $\mathbf{P}Q$ is a one-point *compactification* of E and in this sense the conformal closure is also called the *conformal compactification.*

The notion of conformal compactification can be defined for any space $E_{r,s}$ with $r + s > 2$ and plays a key role in *conformal field theory* (cf. [85]). For a GA account of this process, see [97]. □

The next observation shows two ways of using the conformal vectors of two points to determine their distance.

2.1.7 *Let* $x = H(\boldsymbol{x})$, $x' = H(\boldsymbol{x}')$, $\boldsymbol{x}, \boldsymbol{x}' \in E$. *Then* (1) $(x - x')^2 = (\boldsymbol{x} - \boldsymbol{x}')^2$, *and* (2) $x \cdot x' = -\frac{1}{2}(\boldsymbol{x} - \boldsymbol{x}')^2$.

Proof

(1) We have $x - x' = \boldsymbol{x} - \boldsymbol{x}' + (\kappa - \kappa')e_\infty$, where $\kappa = \boldsymbol{x}^2/2$ and $\kappa' = \boldsymbol{x}'^2/2$. Since e_∞ is null and orthogonal to E, the conclusion is immediate.

(2) Indeed, $-2x \cdot x' = (x - x')^2 - x^2 - x'^2$ and the conclusion follows from (1) and the fact that x and x' are null. □

As already indicated, the orthogonal projection of the null vector a to E is $\boldsymbol{a}$. The proposition above tells us that we can achieve the same by appealing only to the inner product of C:

$$\{\boldsymbol{a}\} = \{\boldsymbol{x} \in E : a \cdot x = 0\}.$$

Indeed, $a \cdot x = 0$ is equivalent to $(\boldsymbol{a} - \boldsymbol{x})^2 = 0$ and hence to $\boldsymbol{x} = \boldsymbol{a}$.

This points out to the possibility of determining geometric objects in E by means of operations in C involving conformal vectors related to the object. Before exploring this possibility in the next two sections, another simple example may help seeing the potential of this idea.

2.1.8 (Example: The bisector plane of a segment) Let $\boldsymbol{a}, \boldsymbol{b} \in E$. Then a point $\boldsymbol{x}$ lies on the bisector plane of the segment $[\boldsymbol{a}, \boldsymbol{b}]$ if and only if $(a - b) \cdot x = 0$. Indeed,

$$\begin{aligned}(a - b) \cdot x &= (\boldsymbol{a} - \boldsymbol{b} + \tfrac{1}{2}(\boldsymbol{a}^2 - \boldsymbol{b}^2)e_\infty) \cdot (\boldsymbol{x} + e_0 + \tfrac{1}{2}\boldsymbol{x}^2 e_\infty) \\ &= (\boldsymbol{a} - \boldsymbol{b}) \cdot \boldsymbol{x} - \tfrac{1}{2}(\boldsymbol{a}^2 - \boldsymbol{b}^2) \\ &= (\boldsymbol{a} - \boldsymbol{b}) \cdot \left(\boldsymbol{x} - \tfrac{1}{2}(\boldsymbol{a} + \boldsymbol{b})\right)\end{aligned}$$

and so the relation $(a - b) \cdot x = 0$ is equivalent to say that $\boldsymbol{x} = \frac{1}{2}(\boldsymbol{a} + \boldsymbol{b}) + \boldsymbol{v}$, where $\boldsymbol{v} \in (\boldsymbol{a} - \boldsymbol{b})^\perp$, and this proves the claim. □

Instead of using the inner product, as in the examples above, we could try to base the representation on the outer product. For example, for a given point $\boldsymbol{a}$, the relation $x \wedge a = 0$ is satisfied by the conformal vectors proportional to a, and all these vectors are conformal representations of $\boldsymbol{a}$ (but only one is normalized). This idea leads to what we will call *outer* representations and will be studied in Sect. 2.3. The representations based on the inner product will be called *inner representations*. We turn to them in the next section.

Caviat. In the literature, outer (inner) representations are called *direct* (*dual*) representations, or OPNS for *outer product null space* (IPNS for *inner product null space*) representations. If it is true that inner and outer representations are dual in a sense that will be made precise later (see 2.3.3), the choosing of which one is *primal* (or direct) and which one is *dual* is a matter of convenience and we prefer to use the terms inner and outer because they refer directly to what is the operation involved.

2.2 Inner Representations

We will consider in turn the inner representations of spheres, planes, circles, and lines. Point pairs will be considered in Sect. 2.3.

We say that $K \in C$ is (or gives) an *inner representation* of $F \subseteq E$ if (and only if)

$$F = \{\boldsymbol{x} \in E : x \cdot K = 0\}.$$

Note that a scalar multiple of an inner representation of F is also an inner representation of F.

Spheres

Consider the sphere $S = S_{z,\rho}$ with center at $\boldsymbol{z} \in E$ and radius ρ. Its points $\boldsymbol{x}$ are those satisfying the equation $(\boldsymbol{x} - \boldsymbol{z})^2 = \rho^2$, which is equivalent to

$$x \cdot z = -\tfrac{1}{2}\rho^2 = x \cdot (\tfrac{1}{2}\rho^2 e_\infty), \quad \text{or} \quad x \cdot (z - \tfrac{1}{2}\rho^2 e_\infty) = 0.$$

2.2.1 (Inner representation of spheres) (1) *The vector*

$$s = z - \tfrac{1}{2}\rho^2 e_\infty = \boldsymbol{z} + e_0 + \tfrac{1}{2}(\boldsymbol{z}^2 - \rho^2)e_\infty$$

is an inner representation of S. Since $s \cdot e_\infty = -1$, we say that s is a normalized *inner representation of S.* (2) *If $\boldsymbol{a} \in E$ is any point on S, $s = a \cdot (z \wedge e_\infty)$.* (3) *The radius and the center of the sphere can be retrieved from s:*

$$s \cdot s = \rho^2, \qquad z = s + \tfrac{1}{2}\rho^2 e_\infty. \tag{2.10}$$

Proof (1) It has already been established.
(2) Since $a \cdot z = -\frac{1}{2}(\boldsymbol{a} - \boldsymbol{z})^2 = -\frac{1}{2}\rho^2$ and $a \cdot e_\infty = -1$, we have

$$s = z - \tfrac{1}{2}\rho^2 e_\infty = z + (a \cdot z)e_\infty = (a \cdot z)e_\infty - (a \cdot e_\infty)z = a \cdot (z \wedge e_\infty).$$

(3) It is a direct consequence of $z^2 = 0$ and $z \cdot e_\infty = -1$. □

In particular we see that points can be construed as spheres of zero radius. By analogy, vectors of the form $s = z + \frac{1}{2}\rho^2 e_\infty$ will be said to be (normalized inner representations of) *imaginary spheres*, and in this case $\rho^2 = -s \cdot s$.

By definition, the quantity $s \cdot x$ vanishes for any point $\boldsymbol{x}$ on the sphere, but it has also meaningful information if it is non-zero.

2.2.2 (Interior and exterior points) *A point $\boldsymbol{x}$ such that $s \cdot x > 0$ ($s \cdot x < 0$) lies in the interior (exterior) of the sphere.*

Proof A short computation shows that $2s \cdot x = \rho^2 - (\boldsymbol{x} - \boldsymbol{z})^2$. If $\boldsymbol{x}$ is interior (exterior) to the sphere, then $(\boldsymbol{x} - \boldsymbol{z})^2 < \rho^2$ ($(\boldsymbol{x} - \boldsymbol{z})^2 > \rho^2$) and $s \cdot x > 0$ ($s \cdot x < 0$). □

Planes

Let $P = P_{\boldsymbol{u},\delta}$ be the plane normal to the unit vector $\boldsymbol{u}$ and with a (signed) perpendicular distance δ to the origin (this means that $\delta\boldsymbol{u} \in P$). Then the points $\boldsymbol{x}$ of P are precisely those that satisfy $\boldsymbol{u} \cdot \boldsymbol{x} = \delta$ (note that the points of P have the form $\boldsymbol{x} = \delta\boldsymbol{u} + \boldsymbol{v}$, with $\boldsymbol{v} \in \boldsymbol{u}^\perp$). Now $\boldsymbol{u} \cdot \boldsymbol{x} = \boldsymbol{u} \cdot x$ and $\delta = -\delta(x \cdot e_\infty)$, so the equation can be written as

$$x \cdot (\boldsymbol{u} + \delta e_\infty) = 0.$$

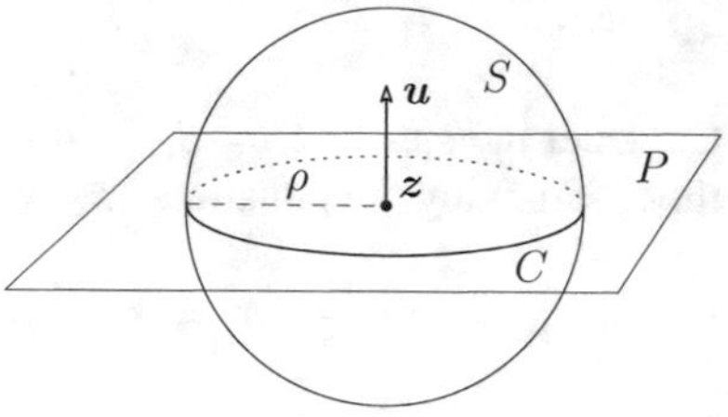

Fig. 2.3 P is the plane through z with unit normal vector $\boldsymbol{u}$, C is the circle on P with center z and radius ρ. If S is the sphere with center z and radius ρ, then P is a diametric plane of S and $C = S \cap P$

2.2.3 (Inner representation of planes) (1) *The vector* $p = \boldsymbol{u} + \delta e_\infty$ *is an inner representation of the plane* P. (2) *For any* $\boldsymbol{a} \in P$, $p \sim a \cdot (\boldsymbol{u} \wedge e_\infty)$ *and hence* $a \cdot (\boldsymbol{u} \wedge e_\infty)$ *is also an inner representation of* P. (3) *Deconstruction:* $\delta = -p \cdot e_0$, $\boldsymbol{u} = p - \delta e_\infty$.

Proof

(1) It has already been proved and (3) is obvious.
(2) If the plane represented by $\boldsymbol{u} + \delta e_\infty$ passes through $\boldsymbol{a}$, then $a \cdot (\boldsymbol{u} + \delta e_\infty) = 0$. So $\delta = (a \cdot \boldsymbol{u})/(-a \cdot e_\infty)$ and $-(a \cdot e_\infty)(\boldsymbol{u} + \delta e_\infty) = -(a \cdot e_\infty)\boldsymbol{u} + (a \cdot \boldsymbol{u})e_\infty = a \cdot (\boldsymbol{u} \wedge e_\infty)$ is also an inner representation of P. □

Circles

We seek an inner representation of the circle C of radius ρ and center z on the plane through z with unit normal vector $\boldsymbol{u}$ (see Fig. 2.3).

2.2.4 *Let* s *be an inner representation of the sphere* $S = S_{z,\rho}$ *and* p *an inner representation of the plane* $P = P_{z,\boldsymbol{u}}$ *through* z *with normal unit vector* $\boldsymbol{u}$. *Then* $c = s \wedge p$ *is an inner representation of* $C = S \cap P$.

Proof By 2.2.1 and 2.2.3, we may assume that $s = z - \frac{1}{2}\rho^2 e_\infty$ and $p = z \cdot (\boldsymbol{u} \wedge e_\infty)$. We want to show that $x \cdot (s \wedge p) = 0$ is equivalent to say that $\boldsymbol{x} \in S \cap P$.

Since $x \cdot (s \wedge p) = (x \cdot s)p - (x \cdot p)s$, this expression vanishes if $\boldsymbol{x}$ lies on the circle, because then $x \cdot s = 0$ (as $\boldsymbol{x}$ belongs to the sphere) and $x \cdot p = 0$ (as $\boldsymbol{x}$ belongs to the plane).

Conversely, assume $(x \cdot s)p - (x \cdot p)s = 0$. Then the dot product with s gives

$$(x \cdot s)(s \cdot p) - (x \cdot p)(s \cdot s) = 0.$$

In this expression, $s \cdot p = (z - \frac{1}{2}\rho^2 e_\infty) \cdot (z \cdot (\boldsymbol{u} \wedge e_\infty))$ turns out to vanish because both $z \cdot (z \cdot (\boldsymbol{u} \wedge e_\infty))$ and $e_\infty \cdot (z \cdot (\boldsymbol{u} \wedge e_\infty))$ vanish, as it is easily checked. So we have $(x \cdot p)(s \cdot s) = 0$ and hence $x \cdot p = 0$ because $s \cdot s = \rho^2 \neq 0$. Now we also have $(x \cdot s)p = 0$, and so $x \cdot s = 0$. Thus $\boldsymbol{x}$ belongs to the circle because it belongs to the sphere and to the plane. □

2.2.5 (Retrieving the radius of C) $\rho^2 = -c^2$.

Proof We know that $s \cdot s = \rho^2$. We also have $p \cdot p = \boldsymbol{u} \cdot \boldsymbol{u} = 1$. On the other hand, $s \cdot p = (z - \frac{1}{2}\rho^2 e_\infty) \cdot (\boldsymbol{u} + \delta e_\infty) = z \cdot (\boldsymbol{u} + \delta e_\infty) = z \cdot \boldsymbol{u} - \delta = 0$ (as $\boldsymbol{z}$ lies on P) and so $c^2 = (s \wedge p) \cdot (s \wedge p) = -\big((s \cdot s)(p \cdot p) - (s \cdot p)^2\big) = -\rho^2$. □

For an expression for the center of C, see E.2.7, p. 51.

Lines

Here we will write $\boldsymbol{u}^*$ ($\boldsymbol{u}^\perp$) for the dual (perpendicular plane) of $\boldsymbol{u} \in E$ in $\mathcal{G}_3$.

2.2.6 (Line through $\boldsymbol{a}$ parallel to $\boldsymbol{u}$) *Let $\boldsymbol{a}, \boldsymbol{u} \in E$, with $\boldsymbol{u}$ unitary. Then the expression $l = a \cdot (\boldsymbol{u}^* e_\infty)$ provides an inner representation of the line L through $\boldsymbol{a}$ parallel to $\boldsymbol{u}$.*

Proof We know that there is an orthonormal basis $\{\boldsymbol{u}_1, \boldsymbol{u}_2\}$ of $\boldsymbol{u}^\perp$ such that $\boldsymbol{u}^* = \boldsymbol{u}_1\boldsymbol{u}_2$. Then $l = a \cdot (\boldsymbol{u}_1\boldsymbol{u}_2 e_\infty) = a_1\boldsymbol{u}_2 e_\infty - a_2\boldsymbol{u}_1 e_\infty - \boldsymbol{u}_1\boldsymbol{u}_2$, where $a_k = a \cdot \boldsymbol{u}_k$, $k = 1, 2$ (the last sign comes from $a \cdot e_\infty = -1$). Now for a point x we can compute $x \cdot l$ and we find $(a_1x_2 - a_2x_1)e_\infty + (x_2 - a_2)\boldsymbol{u}_1 - (x_1 - a_1)\boldsymbol{u}_2$, with $x_k = x \cdot \boldsymbol{u}_k$, $k = 1, 2$. This expression vanishes if and only if $x_1 = a_1$ and $x_2 = a_2$, which is equivalent to say that $\boldsymbol{x} - \boldsymbol{a} \in \langle \boldsymbol{u}_1, \boldsymbol{u}_2\rangle^\perp = \langle \boldsymbol{u} \rangle$, or $\boldsymbol{x} = \boldsymbol{a} + \lambda \boldsymbol{u}$ ($\lambda \in \mathbb{R}$), and this is the parametric equation of L. □

2.3 Outer Representations

An element K of $\mathcal{C}$ is an *outer representation* of a set $F \subseteq E$ if the relation $x \wedge K = 0$ is equivalent to say that $\boldsymbol{x} \in F$. By what we saw in Sect. 2.1, a point is an outer representation of itself, for $x \wedge a = 0$ is equivalent to say that $x \sim a$ and hence to $\boldsymbol{x} = \boldsymbol{a}$.

2.3.1 (Point pair) (1) *If $\boldsymbol{a}, \boldsymbol{b} \in E$ and $\boldsymbol{a} \neq \boldsymbol{b}$, then $a \wedge b$ is an outer representation of the point pair $\{\boldsymbol{a}, \boldsymbol{b}\}$.* (2) $(a \wedge b)^2 = (\boldsymbol{a} - \boldsymbol{b})^4/4$.

Proof (1) The relation $x \wedge a \wedge b = 0$ is equivalent to $x \in \langle a, b\rangle$, or

$$x = \lambda a + \mu b \sim \lambda \boldsymbol{a} + (1 - \lambda)\boldsymbol{b} + e_0 + \frac{1}{2}(\lambda \boldsymbol{a}^2 + (1 - \lambda)\boldsymbol{b}^2)e_\infty$$

(normalize with $\lambda + \mu = 1$). Imposing that x is null, we get $\lambda(1-\lambda)(\boldsymbol{a}-\boldsymbol{b})^2 = 0$, whose solutions are $\lambda = 0$ and $\lambda = 1$.

(2) This follows from $a^2 = b^2 = 0$, $a \cdot b = -\frac{1}{2}(\boldsymbol{a} - \boldsymbol{b})^2$, and $(a \wedge b)^2 = (a \cdot b)^2 = \left(\frac{1}{2}(\boldsymbol{a} - \boldsymbol{b})^2\right)^2$. □

2.3.2 (Line through two points) *Let* $\boldsymbol{a}, \boldsymbol{b} \in E$, $\boldsymbol{a} \neq \boldsymbol{b}$. *Then* (1) $L = a \wedge b \wedge e_\infty$ *is an outer representation of the line joining* $\boldsymbol{a}$ *and* $\boldsymbol{b}$, *and* (2) $L^2 = (\boldsymbol{a} - \boldsymbol{b})^2$.

Proof (1) Indeed, $x \wedge L = 0$ is equivalent to say that $x \in \langle a, b, e_\infty \rangle$:

$$x = \lambda a + \mu b + \xi e_\infty = \lambda \boldsymbol{a} + \mu \boldsymbol{b} + (\lambda + \mu) e_0 + \tfrac{1}{2}(\lambda \boldsymbol{a}^2 + \mu \boldsymbol{b}^2 + 2\xi) e_\infty.$$

Since we can normalize x, we can assume that $\lambda + \mu = 1$, or $\mu = 1 - \lambda$, and so

$$x = \lambda \boldsymbol{a} + (1 - \lambda)\boldsymbol{b} + e_0 + \tfrac{1}{2}\left(\lambda \boldsymbol{a}^2 + (1 - \lambda)\boldsymbol{b}^2 + 2\xi\right) e_\infty.$$

Now it is immediate that there is a unique value of ξ for which x is null and so we conclude that in E we have $\boldsymbol{x} = \lambda \boldsymbol{a} + (1 - \lambda)\boldsymbol{b}$, $\lambda \in \mathbb{R}$, which are precisely the points on the line determined by $\boldsymbol{a}$ and $\boldsymbol{b}$.

(2) $L^2 = -q(a \wedge b \wedge e_\infty)$. Using that $a^2 = b^2 = e_\infty^2 = 0$, $a \cdot b = -\frac{1}{2}(\boldsymbol{a} - \boldsymbol{b})^2$, and $a \cdot e_\infty = b \cdot e_\infty = -1$, the claim follows from Gram's formula. □

Before exploring the outer representations of circles, planes, and spheres, it is convenient to establish a *duality theorem* that relates the inner and outer representations. For this we need the pseudoscalar $\mathcal{I} = e_1 \wedge e_2 \wedge e_3 \wedge e_0 \wedge e_\infty$ and the Hodge duality notation $A^* = A\mathcal{I}$. Note that we have $A^{**} \sim A$, because $\mathcal{I}^2 = -1$ (cf. Fig. 2.1, p. 35). It is also important to note that if A is a k-blade then A^* is a $(5 - k)$-blade (E.2.4, p. 51).

2.3.3 (Duality) *Let* A *be a* k*-vector* $(1 \leqslant k \leqslant 4)$. *Then for any conformal vector* x *we have* $(x \wedge A)^* = x \cdot A^*$. *In particular we have that the relations* $x \wedge A = 0$ *and* $x \cdot A^* = 0$ *are equivalent, which means that the outer representation of* A *agrees with the inner representation of* A^*. *Conversely, replacing* A *by* A^* *we see that the inner representation of* A *agrees with the outer representation of* A^*.

Proof The expressions $(x \wedge A)^*$ and $x \cdot A^*$ are bilinear in x and A, so it suffices to see that they are equal for $x = e_j$, $A = e_K$. If $j \in K$, $e_j \wedge e_K = 0$ and $e_j \cdot e_K^*$ also vanishes because $e_K^* = e_K \mathcal{I}$ does not contain e_j (it is absorbed by $\mathcal{I}$). And if $j \notin K$, $(e_j \wedge e_K)\mathcal{I} = e_j e_K \mathcal{I} = e_j(e_K^*) = e_j \cdot (e_K^*)$, the latter step because e_j *appears* in e_K^*. □

2.3.4 (Circle through three non-collinear points) *If* $\boldsymbol{a}_1, \boldsymbol{a}_2, \boldsymbol{a}_3 \in E$ *are three non-collinear points, then* (1) $C = a_1 \wedge a_2 \wedge a_3$ *is an outer representation of the circle* K *determined by* $\boldsymbol{a}_1, \boldsymbol{a}_2, \boldsymbol{a}_3$. (2) $C^2 = 4\rho^2\Delta^2$, *where* ρ *is the radius of* K *and* Δ *the area of the triangle* $\boldsymbol{a}_1\boldsymbol{a}_2\boldsymbol{a}_3$.

Proof

(1) Let $B \in \mathcal{C}^2$ be a 2-blade whose inner representation is K (see 2.2.4). Then the outer representation of the 3-blade B^* is also K. It is therefore enough to show that $B^* \sim C$. For this, we can argue as follows. The space $V = |B^*\rangle$ of vectors such that $x \wedge B^* = 0$ has dimension 3, and it clearly contains a_1, a_2, a_3, so $V = \langle a_1, a_2, a_3 \rangle$. The claim now follows from $B^*, C \in \wedge^3 V$.

(2) We have $C^2 = -q(a_1 \wedge a_2 \wedge a_3)$. Since $a_1^2 = a_2^2 = a_3^2 = 0$ and

$$a_j \cdot a_k = -\tfrac{1}{2}d_l^2,\ d_l^2 = (\boldsymbol{a}_j - \boldsymbol{a}_k)^2,\ jkl \text{ a cyclic permutation of } 123,$$

we easily obtain $C^2 = \frac{1}{4}d_1^2 d_2^2 d_3^2$. To conclude, it is enough to remember the classical formula $4\rho\Delta = d_1 d_2 d_3$ (cf. [22, Exercise 5, p. 60]). □

Remark that the expression $L = a \wedge b \wedge e_\infty$ for the line defined by $\boldsymbol{a}$ and $\boldsymbol{b}$ can now be phrased by saying that *lines are circles through infinity*.

2.3.5 (Plane through three non-collinear points) (1) *If $\boldsymbol{a}, \boldsymbol{v}, \boldsymbol{w} \in E$ and $\boldsymbol{v}, \boldsymbol{w}$ are linearly independent, then the outer representation of $P = a \wedge \boldsymbol{v} \wedge \boldsymbol{w} \wedge e_\infty$ is the plane through $\boldsymbol{a}$ directed by $\boldsymbol{v} \wedge \boldsymbol{w}$.* (2) *If $\boldsymbol{a}, \boldsymbol{b}, \boldsymbol{c}$ are three non-collinear points, then the outer representation of $P = a \wedge b \wedge c \wedge e_\infty$ is the plane determined by $\boldsymbol{a}, \boldsymbol{b}, \boldsymbol{c}$.* (3) *$P^2 = -16\Delta^2$, where Δ is the area of the triangle $\boldsymbol{abc}$.*

Proof

(1) We have $a \wedge \boldsymbol{v} \wedge \boldsymbol{w} \wedge e_\infty = (\boldsymbol{a} + e_0) \wedge \boldsymbol{v} \wedge \boldsymbol{w} \wedge e_\infty$, for the factor e_∞ kills the e_∞ part of a. Thus the normalized vectors x such that $x \wedge P = 0$ have the form

$$x = \boldsymbol{a} + e_0 + \lambda\boldsymbol{v} + \mu\boldsymbol{w} + \nu e_\infty, \quad \lambda, \mu, \nu \in \mathbb{R}.$$

If we further impose that x is null, we simply have to set $\nu = \frac{1}{2}\boldsymbol{x}^2$, with $\boldsymbol{x} = \boldsymbol{a} + \lambda\boldsymbol{v} + \mu\boldsymbol{w}$ ($\lambda, \mu \in \mathbb{R}$), which is the parametric expression of the points of the plane trough $\boldsymbol{a}$ directed by $\boldsymbol{v} \wedge \boldsymbol{w}$.

(2) We can write $P = a \wedge (b - a) \wedge (c - a) \wedge e_\infty$. As in (1), the factor e_∞ kills the e_∞ part of the other factors and so $P = a \wedge (\boldsymbol{b} - \boldsymbol{a}) \wedge (\boldsymbol{c} - \boldsymbol{a}) \wedge e_\infty$. By (1), the outer representation of P is the plane through $\boldsymbol{a}$ directed by $(\boldsymbol{b} - \boldsymbol{a}) \wedge (\boldsymbol{c} - \boldsymbol{a})$, and this is the plane $\boldsymbol{abc}$.

(3) We have $P^2 = q(a \wedge b \wedge c \wedge e_\infty)$, which can be evaluated with the corresponding Gram determinant with $a^2 = b^2 = c^2 = 0$ and, with obvious meaning of the notations, $a \cdot b = -\frac{1}{2}d_3^2$, $a \cdot c = -\frac{1}{2}d_2^2$, $b \cdot c = -\frac{1}{2}d_1^2$, $a \cdot e_\infty = b \cdot e_\infty = c \cdot e_\infty = -1$. The result is identified as $-16\Delta^2$ by means of Heron's classical formula [22, p. 58]:

$$\begin{aligned} 16\Delta^2 &= (d_1 + d_2 + d_3)(-d_1 + d_2 + d_3)(d_1 - d_2 + d_3)(d_1 + d_2 - d_3) \\ &= (d_1^2 + d_2^2 + d_3^2)^2 - 2(d_1^4 + d_2^4 + d_3^4). \end{aligned}$$
□

2.3.6 (Sphere through four points) (1) *If $\boldsymbol{a}, \boldsymbol{b}, \boldsymbol{c}, \boldsymbol{d} \in E$ are non-coplanar, then the outer representation of $S = a \wedge b \wedge c \wedge d$ is the sphere determined by the four points.* (2) *The radius ρ of the sphere is determined by $\rho^2 = S^2/(S \wedge e_\infty)^2$.*

Proof

(1) It is similar to the proof of 2.3.4. Let $s \in C^1$ be such that its inner representation is the sphere through our four points [see 2.2.1 and Eq. (2.10)]. Then the outer representation of s^* is the same sphere and so we only need to see that $s^* \sim S$,

or, equivalently, that $|s^*\rangle = |S\rangle$. Now $|S\rangle = \langle a, b, c, d\rangle$ and $|s^*\rangle = \{x \in \bar{E} : x \wedge s^* = 0\} = \{x \in \bar{E} : x \cdot s = 0\}$, which must coincide with $|S\rangle$ because both spaces have dimension 4 and $a \cdot s = b \cdot s = c \cdot s = d \cdot s = 0$.

(2) If $S = \alpha s^*$, $\alpha \in \mathbb{R}$, then $S^2 = \alpha^2 (s^*)^2$. But $(s^*)^2 = (s\mathcal{I})^2 = q(s\mathcal{I}) = s\mathcal{I}\tilde{\mathcal{I}}s = -s^2 = -\rho^2$, so that $S^2 = -\alpha^2\rho^2$ (we assume that s is normalized so that $s^2 = \rho^2$). Finally $\alpha = -e_\infty \cdot (\alpha s) = e_\infty \cdot S^* = (e_\infty \wedge S)^*$ and consequently

$$\alpha^2 = ((e_\infty \wedge S)^*)^2 = -(e_\infty \wedge S)^2 = -(S \wedge e_\infty)^2. \qquad \square$$

Note that the planes $a \wedge b \wedge c \wedge e_\infty$ can be said to be spheres passing through $d = e_\infty$.

2.4 Transformations

Let $\bar{\mathrm{O}} = \mathrm{O}(\bar{E})$ be the isometry group of $\bar{E}$. This group preserves Q and hence we have, for $f \in \bar{\mathrm{O}}$, a well-defined map $Q \to Q$, $x \mapsto x' = f(x)$. In terms of the normalized form $x = \boldsymbol{x} + e_0 + \frac{1}{2}\boldsymbol{x}^2 e_\infty$ (cf. 2.1.4), we can write, if $\lambda = -x' \cdot e_\infty \neq 0$ (equivalent to $x' \in Q'$),

$$x' = \lambda(\boldsymbol{x}' + e_0 + \mu e_\infty), \quad \boldsymbol{x}' \in E.$$

Since x' is null, we actually have $\mu = \frac{1}{2}\boldsymbol{x}'^2$, or $x' \sim H(\boldsymbol{x}')$, and the map $E \to E$, $\boldsymbol{x} \mapsto \boldsymbol{x}'$ will be denoted by $\boldsymbol{f}$. We will say that it is the *geometric transformation* associated with f. Thus we have:

$$f(H(\boldsymbol{x})) \sim H(\boldsymbol{f}(\boldsymbol{x})). \tag{2.11}$$

Note also that $\boldsymbol{f}(\boldsymbol{x})$ is undefined when $x' \sim e_\infty$, which happens at most for one point (as we will see, this exception only occurs for inversions). Note also that if x is fixed by f (so $x' = x$), then $\boldsymbol{x}' = \boldsymbol{x}$ and so $\boldsymbol{x}$ is fixed by $\boldsymbol{f}$.

2.4.1 (Composition of geometric transformations) *If $f, g \in \bar{\mathrm{O}}$, then the composition $\boldsymbol{g}\boldsymbol{f}$ is the geometric transformation associated with the composition gf.*

Proof Applying g to $f(H(\boldsymbol{x})) \sim H(\boldsymbol{f}(\boldsymbol{x}))$ we get $g(f(H(\boldsymbol{x}))) \sim g(H(\boldsymbol{f}(\boldsymbol{x})))$ and $g(H(\boldsymbol{f}(\boldsymbol{x}))) \sim H(\boldsymbol{g}(\boldsymbol{f}(\boldsymbol{x})))$. If we write $h = gf$, then $g(f(H(\boldsymbol{x}))) \sim H(\boldsymbol{h}(\boldsymbol{x}))$ and so $H(\boldsymbol{h}(\boldsymbol{x})) \sim H(\boldsymbol{g}(\boldsymbol{f}\boldsymbol{x})))$. Since both sides of this relation are normalized, we conclude that $\boldsymbol{h}(\boldsymbol{x}) = \boldsymbol{g}(\boldsymbol{f}(\boldsymbol{x}))$. $\square$

2.4.2 (The isometry group O is a subgroup of $\bar{\mathrm{O}}$) *The group* O *of isometries of E can be identified with the subgroup of $\bar{\mathrm{O}}$ formed by the isometries f such that $f(e_0) = e_0$ and $f(e_\infty) = e_\infty$. For such an f, $\boldsymbol{f}(\boldsymbol{x}) = f(\boldsymbol{x})$.*

Proof Given $f \in \mathrm{O}$, we can consider its extension $\bar{f} \in \bar{\mathrm{O}}$ which is the identity on $\langle e_0, e_\infty \rangle$. This defines a monomorphism $\mathrm{O} \to \bar{\mathrm{O}}$, $f \mapsto \bar{f}$. Conversely, if $g \in \bar{\mathrm{O}}$ is the identity on $\langle e_0, e_\infty \rangle$, then it leaves invariant $E = \langle e_0, e_\infty \rangle^\perp$ and $g = \bar{f}$ for $f = g|_E$. Finally, $H(\boldsymbol{f}(\boldsymbol{x})) \sim f(H(\boldsymbol{x}))$ and in this case $f(H(\boldsymbol{x})) = H(f(\boldsymbol{x}))$, so $\boldsymbol{f}(\boldsymbol{x}) = f(\boldsymbol{x})$. □

Now we can associate an isometry $\underline{R} \in \bar{\mathrm{O}}$ with any *rotor* $R \in C^+$ (recall that this means that $R\tilde{R} = 1$): $\underline{R}(x) = Rx\tilde{R}$. It is indeed an isometry, for $(Rx\tilde{R})^2 = Rx\tilde{R}Rx\tilde{R} = x^2$.

2.4.3 (Geometric covariance) *The isometry $\underline{R}$ is an automorphism of C in the strong sense that it preserves the geometric, outer and inner products, and is compatible with the involutions.*

Proof Since $\underline{R}(xy) = R(xy)\tilde{R} = Rx\tilde{R}Ry\tilde{R} = \underline{R}(x)\underline{R}(y)$, $\underline{R}$ is an automorphism of the geometric product. From the definitions of the outer and inner products in terms of the geometric product, it is also clear that $\underline{R}$ is an automorphism of these products. Finally, $\widehat{Rx\tilde{R}} = R\hat{x}\tilde{R}$ because R is an even multivector and $\widetilde{Rx\tilde{R}} = \tilde{\tilde{R}}\tilde{x}\tilde{R} = R\tilde{x}\tilde{R}$. □

Now we will analyze in turn the kinds of isometries we get when we select different kinds of rotors and also the geometric transformations they induce.

2.4.4 (Rotations) We know that the rotation of $\boldsymbol{x} \in E$ about the origin by an angle $\boldsymbol{i}\theta$ *($\boldsymbol{i}$ a unit area and $\theta \in \mathbb{R}$)* is given by the rotor $R = e^{-\boldsymbol{i}\theta/2}$. In the conformal space, this rotation is given by the same formula (see 2.4.2):

$$\underline{R}(x) = e^{-\boldsymbol{i}\theta/2} x e^{\boldsymbol{i}\theta/2}. \tag{2.12}$$

Proof A direct consequence of the fact that the unit bivector $\boldsymbol{i} \in \mathcal{G}_3^2$ commutes with e_0 and e_∞ and hence these vectors are fixed by $\underline{R}$. □

The transformation (2.12) will be denoted by $\mathsf{R}_{\boldsymbol{i}\theta}$ and in this case, $\boldsymbol{\mathsf{R}}_{\boldsymbol{i}\theta}(\boldsymbol{x}) = \mathsf{R}_{\boldsymbol{i}\theta}(\boldsymbol{x})$.

In what follows, the quadratic coefficient $\frac{1}{2}\boldsymbol{x}^2$ of e_∞ in x will be denoted by $\kappa(\boldsymbol{x})$:

$$x = \boldsymbol{x} + e_0 + \kappa(\boldsymbol{x})e_\infty. \tag{2.13}$$

2.4.5 (Translations) *Let $\boldsymbol{v} \in E$ and consider the rotor $R = e^{-\boldsymbol{v}e_\infty/2}$ and the isometry $\mathsf{T}_{\boldsymbol{v}} = \underline{R}$. Then $\mathsf{T}_{\boldsymbol{v}}(e_\infty) = e_\infty$ and $\boldsymbol{\mathsf{T}}_{\boldsymbol{v}}(\boldsymbol{x}) = \boldsymbol{x} + \boldsymbol{v}$* (the translation by $\boldsymbol{v}$).

Proof It consists of several computational steps.

(*a*) Since $(\boldsymbol{v}e_\infty)^2 = 0$, $e^{\pm \boldsymbol{v}e_\infty/2} = 1 \pm \boldsymbol{v}e_\infty/2$ and

$$\mathsf{T}_{\boldsymbol{v}}(x) = \left(1 - \tfrac{1}{2}\boldsymbol{v}e_\infty\right) x \left(1 + \tfrac{1}{2}\boldsymbol{v}e_\infty\right). \tag{2.14}$$

From this, and again $e_\infty^2 = 0$, we see that e_∞ is fixed and hence the part $\kappa(\boldsymbol{x})e_\infty$ of x is not changed by $\boldsymbol{T_v}$.

(*b*) The transformation of e_0 is v:

$$T_{\boldsymbol{v}}(e_0) = v = \boldsymbol{v} + e_0 + \kappa(\boldsymbol{v})e_\infty. \tag{2.15}$$

Here is why:

$$\begin{aligned} T_{\boldsymbol{v}}(e_0) &= \left(1 - \tfrac{1}{2}\boldsymbol{v}e_\infty\right) e_0 \left(1 + \tfrac{1}{2}\boldsymbol{v}e_\infty\right) = \left(e_0 - \tfrac{1}{2}\boldsymbol{v}e_\infty e_0\right)\left(1 + \tfrac{1}{2}\boldsymbol{v}e_\infty\right) \\ &= e_0 + \tfrac{1}{2}e_0\boldsymbol{v}e_\infty - \tfrac{1}{2}\boldsymbol{v}e_\infty e_0 - \tfrac{1}{4}\boldsymbol{v}e_\infty e_0\boldsymbol{v}e_\infty \\ &= e_0 - \tfrac{1}{2}\boldsymbol{v}\left(e_0e_\infty + e_\infty e_0\right) - \tfrac{1}{4}e_\infty e_0 e_\infty \boldsymbol{v}^2. \end{aligned}$$

Now $-\frac{1}{2}\boldsymbol{v}(e_0e_\infty + e_\infty e_0) = -\boldsymbol{v}(e_0 \cdot e_\infty) = \boldsymbol{v}$, and the second summand of the last expression reduces to $\boldsymbol{v}$. In the third summand we have $e_\infty e_0 e_\infty = e_\infty(-e_\infty e_0 + 2(e_0 \cdot e_\infty)) = -2e_\infty$, so $T_{\boldsymbol{v}}(e_0) = e_0 + \boldsymbol{v} + \frac{1}{2}\boldsymbol{v}^2 e_\infty = v$.

(*c*) As for $\boldsymbol{x}$, $T_{\boldsymbol{v}}(\boldsymbol{x}) = \boldsymbol{x} + (\boldsymbol{v} \cdot \boldsymbol{x})e_\infty$. Indeed,

$$\begin{aligned} T_{\boldsymbol{v}}(\boldsymbol{x}) &= \left(1 - \tfrac{1}{2}\boldsymbol{v}e_\infty\right) \boldsymbol{x} \left(1 + \tfrac{1}{2}\boldsymbol{v}e_\infty\right) = \left(\boldsymbol{x} + \tfrac{1}{2}\boldsymbol{v}\boldsymbol{x}e_\infty\right)\left(1 + \tfrac{1}{2}\boldsymbol{v}e_\infty\right) \\ &= \boldsymbol{x} + \tfrac{1}{2}\boldsymbol{x}\boldsymbol{v}e_\infty + \tfrac{1}{2}\boldsymbol{v}\boldsymbol{x}e_\infty = \boldsymbol{x} + (\boldsymbol{v} \cdot \boldsymbol{x})\, e_\infty. \end{aligned}$$

(*d*) Together, (*a*)–(*c*) yield:

$$\begin{aligned} T_{\boldsymbol{v}}(x) &= T_{\boldsymbol{v}}(\boldsymbol{x} + e_0 + \kappa(\boldsymbol{x})e_\infty) \\ &= \boldsymbol{x} + \boldsymbol{v} + e_0 + (\kappa(\boldsymbol{x}) + \kappa(\boldsymbol{v}) + \boldsymbol{v} \cdot \boldsymbol{x})e_\infty \\ &= \boldsymbol{x} + \boldsymbol{v} + e_0 + \kappa(\boldsymbol{x} + \boldsymbol{v})e_\infty \\ &= H(\boldsymbol{x} + \boldsymbol{v}). \end{aligned}$$

Therefore $\boldsymbol{T_v}(\boldsymbol{x}) = \boldsymbol{x} + \boldsymbol{v}$ and this ends the proof. □

Because of the form of the rotor giving $T_{\boldsymbol{v}}$ and the fact that it leaves e_∞ invariant, translations are often said to be "rotations about infinity."

Since rotations and translations generate the group $G^+(E)$ of proper affine transformations that preserve distances (the *proper Euclidean group* of E), we see that $G^+(E)$ is embedded in $\bar{\mathrm{O}}$.

2.4.6 (Dilations) *Let D_α ($\alpha \in \mathbb{R}$) be the isometry associated with the rotor $e^{\alpha Z/2}$, so that $D_\alpha(x) = e^{\alpha Z/2}xe^{-\alpha Z/2}$* (recall that $Z = e_{0\infty}$). *Then $\boldsymbol{D}_\alpha(\boldsymbol{x}) = e^\alpha \boldsymbol{x}$* (we say that *$\boldsymbol{D}_\alpha$ it is the dilation of E by e^α*).

Proof We refer to 2.1.2 and 2.1.3 for the justification of several steps in the computation that follows. Since Z commutes with vectors $\boldsymbol{x}$, $e^{\alpha Z/2}\boldsymbol{x}e^{-\alpha Z/2} = \boldsymbol{x}$ for all $\boldsymbol{x} \in E$. On the other hand (use $Z^2 = 1$ and $e_0 Z = -e_0 Z = e_0$),

$$
\begin{aligned}
e^{\alpha Z/2} e_0 e^{-\alpha Z/2} &= (\cosh\tfrac{\alpha}{2} + Z\sinh\tfrac{\alpha}{2})\, e_0\, (\cosh\tfrac{\alpha}{2} - Z\sinh\tfrac{\alpha}{2}) \\
&= (\cosh\tfrac{\alpha}{2} - \sinh\tfrac{\alpha}{2})(\cosh\tfrac{\alpha}{2} - \sinh\tfrac{\alpha}{2}) e_0 \\
&= e^{-\alpha} e_0,
\end{aligned}
$$

and a similar computation yields $e^{\alpha Z/2} e_\infty e^{-\alpha Z/2} = e^{\alpha} e_\infty$. Finally,

$$D_\alpha(\boldsymbol{x}+e_0+\kappa(\boldsymbol{x})e_\infty)=\boldsymbol{x}+e^{-\alpha}e_0+e^{\alpha}\kappa(\boldsymbol{x})e_\infty=e^{-\alpha}(e^{\alpha}\boldsymbol{x}+e_0+\kappa(e^{\alpha}\boldsymbol{x})e_\infty)\sim H(e^{\alpha}\boldsymbol{x}),$$

and this tells us that $\boldsymbol{D}_\alpha(\boldsymbol{x}) = e^{\alpha}(\boldsymbol{x})$. □

With dilations, the proper Euclidean group $G^+(E)$ is enlarged to the group $\Gamma^+(E)$ of *proper similarity transformations* and the preceding analysis shows that we can regard $\Gamma^+(E)$ as a subgroup of $\bar{\mathrm{O}}$. Thus we have

$$\mathrm{O}^+ \subset G^+ \subset \Gamma^+ \subset \bar{\mathrm{O}}^+.$$

Now we turn to improper isometries. Instead of rotor transformations $\underline{R}(x) = Rx\tilde{R}$, to treat them we need *versor isometries* $V_s : \bar{E} \to \bar{E}$ (where s is a non-null conformal vector), which is defined by the formula $V_s(x) = -sxs^{-1}$. By what we will see next, they are also called *conformal reflections*.

2.4.7 (Properties of V_s) *$V_s(s) = s$ and $V_s(x) = -x$ for $x \in s^\perp$. In other words, V_s is the reflection in conformal space across $s^\perp$. This implies that V_s is an isometry and that it is improper* ($\det V_s = -1$).

Proof We have $V_s(s) = -sss^{-1} = -s$. On the other hand, if $x \in s^\perp$ then it anticommutes with s and $V_s(x) = -sxs^{-1} = xss^{-1} = x$. □

2.4.8 (Remark) *V_s is a* linear *automorphism of C, but not an algebra automorphism: $V_s(xy) = -sxys^{-1} = -(-sxs^{-1})(-sys^{-1}) = -V_s(x)V_s(y)$. On the other hand, $V_s(x)\hat{} = V_s(\hat{x})$ (because s is odd), but $V_s(x)\tilde{} = V_{s^{-1}}(\tilde{x})$.* □

We can now find out the geometric transformation $\boldsymbol{V}_s$ associated with V_s in terms of particular specifications of s.

2.4.9 (Euclidean reflections) *Let $\boldsymbol{u} \in E$ be a unit vector and $\delta \in \mathbb{R}$. Then* (1) *$\boldsymbol{V}_{\boldsymbol{u}}$ is the reflection across the plane $\boldsymbol{u}^\perp$ and* (2) *$\boldsymbol{V}_{\boldsymbol{u}+\delta e_\infty}$ is the reflection across the plane $P_{\boldsymbol{u},\delta}$ with normal vector $\boldsymbol{u}$ and perpendicular (signed) distance δ to $\boldsymbol{0}$.*

Proof

(1) From $V_{\boldsymbol{u}}(x) = -\boldsymbol{u}x\boldsymbol{u}$, we see that $V_{\boldsymbol{u}}$ leaves invariant e_0 and e_∞ and therefore $\boldsymbol{V}_{\boldsymbol{u}}(\boldsymbol{x}) = V_{\boldsymbol{u}}(\boldsymbol{x}) = -\boldsymbol{u}\boldsymbol{x}\boldsymbol{u}$, which is, as we know, the expression of the reflection of $\boldsymbol{x}$ across $\boldsymbol{u}^\perp$.
(2) The key is the relation $V_{u+\delta e_\infty} = T_{\delta u} V_u T_{-\delta u}$, as it implies $\boldsymbol{V}_{\boldsymbol{u}+\delta e_\infty} = \boldsymbol{T}_{\delta \boldsymbol{u}} \boldsymbol{V}_{\boldsymbol{u}} \boldsymbol{T}_{-\delta \boldsymbol{u}}$ and this is the claimed reflection:

$$\boldsymbol{V}_{\boldsymbol{u}+\delta e_\infty}(\boldsymbol{x}) = -\boldsymbol{u}(\boldsymbol{x} - \delta\boldsymbol{u})\boldsymbol{u} + \delta\boldsymbol{u} = -\boldsymbol{u}\boldsymbol{x}\boldsymbol{u} + 2\delta\boldsymbol{u}.$$

To end the proof, note that $(T_{\delta u} V_u T_{-\delta u})(x) = V_p(x)$, where $p = e^{-\delta u e_\infty/2} u e^{\delta u e_\infty/2}$, and then

$$p = (1 - \delta u e_\infty/2) u (1 + \delta u e_\infty/2) = (u + \delta e_\infty/2)(1 + \delta u e_\infty/2) = u + \delta e_\infty.$$

□

With *reflections*, the groups $G^+(E)$ and $\Gamma^+(E)$ are enlarged to the full *Euclidean group* $G(E)$ of distance-preserving affine transformations and to the group $\Gamma(E)$ of similarities, respectively. Thereby we have the following embeddings:

$$\mathrm{O} \subset G \subset \Gamma \subset \bar{\mathrm{O}}.$$

2.4.10 (Inversions) *Let* $s = e_0 - \frac{1}{2}\rho^2 e_\infty$ (recall that its inner representation is the sphere of radius ρ centered at $\mathbf{0} \in E$). *Then the isometry* V_s *satisfies:* (1) $V_s(e_0) = \frac{\rho^2}{2} e_\infty$ *and* $V_s(e_\infty) = \frac{2}{\rho^2} e_0$. (2) $\boldsymbol{V}_s(\boldsymbol{x}) = \rho^2/\boldsymbol{x}$, *which means that* $\boldsymbol{V}_s$ *is the inversion with respect to the sphere of radius* ρ *centered at* $\mathbf{0}$. (3) *For any* $z \in E$, *the geometric transformation associated with the isometry* $V_{z,\rho} = T_z V_s T_{-z}$ *is the inversion with respect to the sphere of radius* ρ *centered at* z.

Proof

(1) Since $s^2 = \rho^2$, $s^{-1} = s/\rho^2$. Then

$$V_s(e_0) = -s e_0 s/\rho^2 = \tfrac{1}{2} e_\infty e_0 (e_0 - \tfrac{\rho^2}{2} e_\infty) = -\tfrac{\rho^2}{4} e_\infty e_0 e_\infty = \tfrac{\rho^2}{2} e_\infty.$$

The proof of the other relation is very similar.

(2) For $x = \boldsymbol{x} + e_0 + \kappa(\boldsymbol{x}) e_\infty$, $V_s(x) = V_s(\boldsymbol{x}) + V_s(e_0) + \kappa(\boldsymbol{x}) V_s(e_\infty)$. Now $V_s(\boldsymbol{x}) = -s\boldsymbol{x}s^{-1} = \boldsymbol{x}$ (as $\boldsymbol{x}$ anticommutes with s), which together with (1) yields $\boldsymbol{V}_s(\boldsymbol{x}) = \rho^2/\boldsymbol{x}$:

$$V_s(x) = \boldsymbol{x} + \tfrac{2}{\rho^2}\kappa(\boldsymbol{x}) e_0 + \tfrac{\rho^2}{2} e_\infty \sim H(\tfrac{\rho^2}{\boldsymbol{x}^2}\boldsymbol{x}) = H(\rho^2/\boldsymbol{x}).$$

(3) By 2.4.1 we have $\boldsymbol{V}_{z,\rho} = \boldsymbol{T}_z \boldsymbol{V}_s \boldsymbol{T}_{-z}$. Thus $\boldsymbol{V}_{z,\rho}(\boldsymbol{x}) = \boldsymbol{V}_s(\boldsymbol{x} - z) + z = \frac{\rho^2}{\boldsymbol{x}-z} + z$, where in the last step we have used (2). □

2.4.11 (Remark on the conformal group) The tools developed in this section are sufficient to prove that all geometric transformations induced by isometries of $\bar{E}$ are *conformal*, which means that they preserve (non-oriented) angles (see E.2.2, p. 50). Conformal transformations of E form a group with the composition, $\mathrm{Conf}(E)$, and so we have a group homomorphism

$$\bar{\mathrm{O}} \to \mathrm{Conf}(E), \; f \mapsto \boldsymbol{f}.$$

Now this homomorphism is surjective as a consequence of a classical theorem of Liouville asserting that any element of Conf(E) is the composition of inversions in spheres and reflexions in planes [70], and we have shown that these belong to the image. Finally, it is straightforward to check, for $f \in \bar{\mathrm{O}}$, that $\boldsymbol{f} = \mathbf{Id}$ if and only if $f = \pm\mathrm{Id}$, which can be stated as $\mathrm{Conf}(E) \simeq \bar{\mathrm{O}}/\{\pm\mathrm{Id}\}$. □

2.5 Exercises

E.2.1 (Distances with conformal algebra)

(1) In 2.1.7 we have seen how to get the squared distance between two points $\boldsymbol{x}, \boldsymbol{x}' \in E$ using the inner product of their conformal vectors x, x':

$$(\boldsymbol{x} - \boldsymbol{x}')^2 = (x - x')^2 = -2x \cdot x'.$$

(2) From 2.2.3 we know that if $\boldsymbol{u}$ is a unit vector and $\delta \in \mathbb{R}$, then $p = \boldsymbol{u} + \delta e_\infty$ is an inner representation of the plane P with normal vector $\boldsymbol{u}$ and signed distance δ to $\mathbf{0}$ (meaning that $\delta\boldsymbol{u} \in P$). We also have $\delta = -p \cdot e_0$. Now it is immediate that $p \cdot x = \boldsymbol{x} \cdot \boldsymbol{u} - \delta$. If we write $\boldsymbol{x} = \delta'\boldsymbol{u} + \boldsymbol{v}$, with $\boldsymbol{v} \in \boldsymbol{u}^\perp$ (so $\delta' = \boldsymbol{x} \cdot \boldsymbol{u}$), we get $p \cdot x = \delta' - \delta$, which is the (signed) distance from $\boldsymbol{x}$ to P (see Fig. 2.4a).

(3) Let $s = z - \frac{1}{2}\rho^2 e_\infty$ be the inner representation of the sphere $S = S_{\boldsymbol{z},\rho}$ of radius ρ and center $\boldsymbol{z}$, where z is the conformal vector of $\boldsymbol{z}$ (see 2.2.1). In the proof of 2.2.2, we have noticed that $-2s \cdot x = \delta^2 - \rho^2$, where $\delta^2 = (\boldsymbol{x} - \boldsymbol{z})^2$ is the squared distance of $\boldsymbol{x}$ to the center $\boldsymbol{z}$. Since $\delta^2 - \rho^2 = (\delta + \rho)(\delta - \rho)$, we conclude that $-2s \cdot x$ is *the power of* $\boldsymbol{x}$ *with respect to* S, and hence that it is the square of the length of any segment $[\boldsymbol{x}\boldsymbol{a}]$ with $\boldsymbol{a} \in S$ such that the line $\boldsymbol{x}\boldsymbol{a}$ is tangent to S (see Fig. 2.4b).

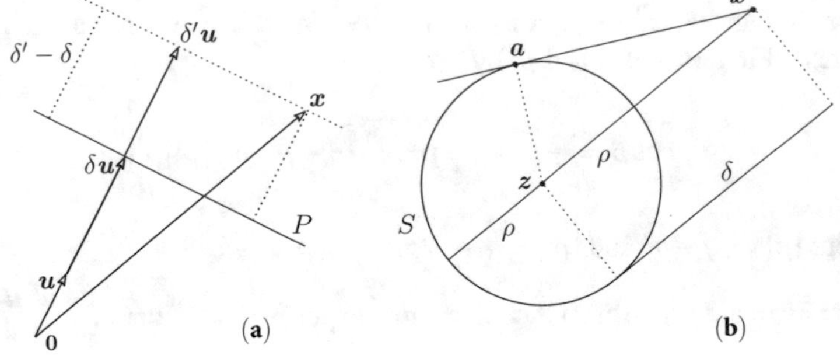

Fig. 2.4 (**a**) Signed distance from a point $\boldsymbol{x}$ to the plane $P = P_{\boldsymbol{u}+\delta e_\infty}$: $\delta' - \delta = p \cdot x$. (**b**) Interpretation of $-2s \cdot x$ as $\delta^2 - \rho^2$, which is the power of $\boldsymbol{x}$ with respect to the sphere $S = S_{\boldsymbol{z},\rho}$

(4) Check that $p \cdot s = \boldsymbol{u} \cdot z - \delta$. By (2), this is the signed distance from z to P. In particular we see that p and s are orthogonal if and only if P passes through the center of z, or, in other words, if P is a diametric plane of S.

(5) Let $s' = z' - \frac{1}{2}\rho'^2 e_\infty$ be the inner representations of $S' = S_{z',\rho'}$. Show that $2s \cdot s' = \rho^2 + \rho'^2 - \delta^2$, where $\delta^2 = (z - z')^2$. In particular we see that s and s' are orthogonal if and only if $\rho^2 + \rho'^2 = \delta^2$.

E.2.2 (Angles with conformal algebra)

(1) Let $L = L_{\boldsymbol{a},\boldsymbol{u}}$ be the line through $\boldsymbol{a} \in E$ with direction $\langle \boldsymbol{u} \rangle$ and let $L' = L_{\boldsymbol{a}',\boldsymbol{u}'}$ be defined similarly. The angle α between these two lines is defined as the angle $\alpha = \alpha(\boldsymbol{u}, \boldsymbol{u}')$ and hence by $\cos\alpha = (\boldsymbol{u} \cdot \boldsymbol{u}')/|\boldsymbol{u}||\boldsymbol{u}'|$. Let $l = a \wedge \boldsymbol{u} \wedge e_\infty$ and $l' = a' \wedge \boldsymbol{u}' \wedge e_\infty$ be the outer representations of L and L'. Show that $l \cdot l' = \boldsymbol{u} \cdot \boldsymbol{u}'$, and in particular that $l^2 = \boldsymbol{u}^2$, so that $|l| = |\boldsymbol{u}|$, and similarly $|l'| = |\boldsymbol{u}'|$. Consequently, we can express α in terms of l and l' by $\cos\alpha = (l \cdot l')/|l||l'|$. *Hint:* Use Gram's formula to evaluate $l \cdot l'$ and notice that the values below the secondary diagonal of Gram's matrix are 0, so that the values above that diagonal are irrelevant.

(2) Let $p = \boldsymbol{u} + \delta e_\infty$ be the inner representation of the plane $P = P_{\boldsymbol{u},\delta}$ with normal unit vector $\boldsymbol{u}$ and signed distance δ to the origin, and let $p' = \boldsymbol{u}' + \delta' e_\infty$ be defined in a similar way. Then it is clear that $p \cdot p' = \boldsymbol{u}\boldsymbol{u}' = \cos\alpha$, where $\alpha = \alpha(\boldsymbol{u}, \boldsymbol{u}')$ agrees with the dihedral angle formed by P and P'.

(3) Let $P = P_{\boldsymbol{a},\boldsymbol{u},\boldsymbol{v}}$ be the plane through $\boldsymbol{a} \in E$ with direction $\langle \boldsymbol{u}, \boldsymbol{v} \rangle$ and let $P' = P_{\boldsymbol{a}',\boldsymbol{u}',\boldsymbol{v}'}$ be defined similarly. The angle β between these two planes is defined as the angle $\beta = \alpha(\boldsymbol{u} \times \boldsymbol{v}, \boldsymbol{u}' \times \boldsymbol{v}')$, that is, as the angle between the normal vectors to the planes. Thus $\cos\beta = ((\boldsymbol{u} \times \boldsymbol{v}) \cdot (\boldsymbol{u}' \times \boldsymbol{v}'))/|\boldsymbol{u} \times \boldsymbol{v}|\,|\boldsymbol{u}' \times \boldsymbol{v}'|$. Now

$$(\boldsymbol{u} \times \boldsymbol{v}) \cdot (\boldsymbol{u}' \times \boldsymbol{v}') = \left((\boldsymbol{u} \wedge \boldsymbol{v})\boldsymbol{i}(\boldsymbol{u}' \wedge \boldsymbol{v}')\boldsymbol{i}\right)_0 = -(\boldsymbol{u} \wedge \boldsymbol{v}) \cdot (\boldsymbol{u}' \wedge \boldsymbol{v}').$$

In particular, $|\boldsymbol{u} \times \boldsymbol{v}|^2 = (\boldsymbol{u} \times \boldsymbol{v})^2 = -(\boldsymbol{u} \wedge \boldsymbol{v})^2$ and $|\boldsymbol{u}' \times \boldsymbol{v}'|^2 = (\boldsymbol{u}' \times \boldsymbol{v}')^2 = -(\boldsymbol{u}' \wedge \boldsymbol{v}')^2$. Let $p = a \wedge \boldsymbol{u} \wedge \boldsymbol{v} \wedge e_\infty$ and $p' = a' \wedge \boldsymbol{u}' \wedge \boldsymbol{v}' \wedge e_\infty$ be the outer representations of P and P'. Show that $p \cdot p' = (\boldsymbol{u} \wedge \boldsymbol{v}) \cdot (\boldsymbol{u}' \wedge \boldsymbol{v}')$, and in particular that $p^2 = (\boldsymbol{u} \wedge \boldsymbol{v})^2$ and $p'^2 = (\boldsymbol{u}' \wedge \boldsymbol{v}')^2$. Consequently, we can express β in terms of p and p' by

$$\cos\beta = -p \cdot p'/\sqrt{p^2 p'^2} = -p \cdot p'/|p||p'|,$$

where $|p| = \sqrt{-p^2}$ and $|p'| = \sqrt{-p'^2}$.

Hint: Evaluate $p \cdot p'$ using Gram's formula to get $p \cdot p' = -\det\begin{pmatrix} \boldsymbol{u} \cdot \boldsymbol{u}' & \boldsymbol{u} \cdot \boldsymbol{v}' \\ \boldsymbol{v} \cdot \boldsymbol{u}' & \boldsymbol{v} \cdot \boldsymbol{v}' \end{pmatrix} = (\boldsymbol{u} \wedge \boldsymbol{v}) \cdot (\boldsymbol{u}' \wedge \boldsymbol{v}')$.

E.2.3 (Rotations about an arbitrary point) (1) The map $\boldsymbol{R}_{z,i\theta} : \boldsymbol{x} \mapsto \boldsymbol{R}_{i\theta}(\boldsymbol{x}-z)+z$ (the *rotation about* z *by* $\boldsymbol{i}\theta$) is the geometrical transformation corresponding to the rotor $T_z R_{i\theta} T_{-z}$. (2) Show that the last expression is equal to $R_{T_z(i\theta)}$.

E.2.4 (The dual of a blade is a blade) Let A be a 2-blade, say $A = a_1 \wedge a_2$. Taking an orthonormal basis in $\langle a_1, a_2\rangle$, we may assume that a_1 and a_2 are part of an orthogonal basis of $\bar{E}$. If $\mathcal{I}$ is the pseudoscalar of this basis, then it is clear that $A^* = (a_1 \wedge a_2)\mathcal{I} = a_1 a_2 \mathcal{I}$ and $a_1 a_2 \mathcal{I}$ is a blade (a_1 and a_2 are contracted with the corresponding factors of $\mathcal{I}$). The same argument works for any k-blade. Note that the space spanned by $A\mathcal{I}$ is the orthogonal of the space spanned by A.

E.2.5 (The center of a sphere revisited) If the inner representation of $s \in C^1$ is a sphere, and s is normalized, we saw how to find its center z (see 2.2.1) and its radius (2.10). Prove that we also have $z \sim s e_\infty s$. This has the advantage that it does not assume that s is normalized.

E.2.6 (Alternative formula for the radius of a circle) With the same notations as in 2.3.4, prove that $\rho^2 = -C^2/(C \wedge e_\infty)^2$. Deduce that $\Delta^2 = -\frac{1}{4}(C \wedge e_\infty)^2$.

E.2.7 (Center of a circle) If the inner representation of $c \in C^2$ is a circle, prove that then its center is $c e_\infty c$.

E.2.8 (Line perpendicular to circle passing through the center) Assume that the inner representation of $c \in C^2$ is a circle C with center at z. Prove that then the inner representation of $l = e_\infty \cdot c$ is the line perpendicular to the plane of C and going through z.

E.2.9 (Further examples about composing transformations) Prove the following statements (we use the notations introduced in Sect. 2.4):

(1) $\boldsymbol{D}_\beta \boldsymbol{D}_\alpha = \boldsymbol{D}_{\alpha+\beta}$.
(2) The map $\boldsymbol{D}_{z,\alpha} : \boldsymbol{x} \mapsto e^\alpha(\boldsymbol{x} - z) + z$ (the *dilation with center* z *by* e^α) is the geometrical transformation corresponding to $T_z D_\alpha T_{-z}$.
(3) The composition of two inversions with respect to concentric spheres is a dilation with respect to that center. Moreover, if ρ and ρ' are the radii of the first and second spheres, then parameter α of the dilation is $2\ln(\rho'/\rho)$.

E.2.10 In conformal space, all similarity transformations (elements of the group $\Gamma(E)$) leave e_∞ invariant (use 2.4.2 and 2.4.4 for isometries of E, 2.4.5 for translations, 2.4.6 for dilations, and 2.4.9 for Euclidean reflections). This implies that similarities map point pairs, circles, lines, spheres, and planes to point pairs, circles, lines, spheres, and planes, respectively (use geometric covariance 2.4.3 and the outer representations of point pairs 2.3.1, circles 2.3.4, lines 2.3.2, spheres 2.3.6, and planes 2.3.5).

On the other hand, in conformal space the inversions exchange e_0 and e_∞ (see 2.4.10). This and geometric covariance imply that inversions map: (1) *flats* (lines and planes) not passing through **0** to *rounds* (circles and spheres) that go through $\mathbf{0} \in E$. (2) rounds through **0** to flats not going through **0**. (3) any flat passing through **0** to itself.

Chapter 3
Minkowski's Space-Time: Geometry and Physics

The purpose of this chapter is a study of Minkowski's space-time that emphasizes the fundamental geometric and physical aspects that concur in its structure.

The language used is linear algebra and its extension to geometric algebra, as presented in Sect. 1.4, which is a prerequisite for Sects. 3.2–3.4. It is the method that appears as best suited for expressing and managing Lorentz transformations, relativistic electrodynamics, and Dirac's electron theory.

As a starting point, we take a *Lorentzian vector space* $E_{1,3}$, that is, a real vector space of dimension 4 with a *metric* (symmetric bilinear form) of signature $(1, 3)$, and the affine space $\mathcal{M}$ (*Minkowski space*) whose associated vector space is $E_{1,3}$. This structure, one of the premises of Minkowski's original article [74], integrates, as we shall see, advances due to names like Maxwell, Lorentz, Poincaré, and Einstein, among others.

Such an approach should not surprise a reader with mathematical training, who is accustomed to take as a starting point for the study of Euclidean geometry an *Euclidean vector space* E_n, that is, a real vector space of dimension n (take $n = 3$ if it is desired to reinforce the analogy) endowed with a positive definite metric, together with the corresponding affine space (*Euclidean space*). In this case the statement synthesizes the understanding of Euclidean geometry achieved over millennia, from the Greeks (Pythagoras, Euclid, Archimedes, etc.), through the "analytical revolution" (Descartes, Pascal, Newton, Euler, etc.), and crystallized with the development of "intrinsic" algebro-geometric structures from seminal contributions such as those of Grassmann and Riemann.

The geometrical and physical aspects of the Minkowski space that do not depend on GA are covered in Sect. 3.1. The remaining three sections offer a GA view of the Lorentz transformations, relativistic electrodynamics, and Dirac's equation. Besides Hestenes' many authoritative works, as, for instance, [40] and [47], and general treatises such as [29], all of which have been an unfailing source of inspiration, more specific references will be given at the places where they are required.

C. Lavor et al., *A Geometric Algebra Invitation to Space-Time Physics, Robotics and Molecular Geometry*, SpringerBriefs in Mathematics,
https://doi.org/10.1007/978-3-319-90665-2_3

3.1 From Physics to Geometry and Back

In this section we will review the main ideas in Minkowski's view of the special theory of relativity and then we will lay out a mathematical presentation of the key geometric facts concerning that space.

Minkowski's View of Special Relativity

The metric proposed by Minkowski is given, using his own terms, by the quadratic form

$$c^2t^2 - (x^2 + y^2 + z^2), \tag{3.1}$$

where x, y, z are rectangular Cartesian coordinates with respect to an inertial reference, t the time relative to it, and c the speed of light in vacuum.

It is crucial to point out that c is a *universal constant*, in the sense that it *does not depend on the inertial reference system in which it is measured nor on the velocity of the emitter focus*. This surprising fact is one of the predictions of Maxwell's theory. Indeed, Maxwell found that the speed of electromagnetic waves in vacuum predicted by his theory should be, regardless of how they are generated, $c = (\epsilon_0\mu_0)^{-1/2}$, where ϵ_0 and μ_0 are physical constants (*electrical permittivity* and *magnetic permeability* of vacuum, respectively) that can be measured in the laboratory and whose values are universal (admitting the *principle of relativity* according to which physical laws have the same form in any inertial system).

The conclusion that light is an electromagnetic wave, thus incorporating *optics* to his theory, was reached by Maxwell when he verified that the numerical value of $(\epsilon_0\mu_0)^{-1/2}$ agreed with the speed of light in vacuum.

In addition to the theoretical prediction, the universality of c has been verified, directly or indirectly, with a variety of experiments ranging from Michelson-Morley's [73] to the sophisticated current GPS systems (cf. [93]).

Let us also say that Einstein took the principle of relativity and the universality of c as axioms in his work [33], thus being able to easily obtain the *Lorentz transformations* which relate the values x, y, z, t relative to an inertial system S with the values x', y', z', t' relative to another inertial system S'. If time is not absolute, these transformations can be deduced without assuming the constancy of c, as done, for example, in [95].

The relevance of the Lorentz metric lies in the fact that the special Lorentz transformation (*Lorentz boost* according to the usual terminology) is an isometry of (3.1). This statement is checked with a simple calculation using the expression of the boost. In units such that $c = 1$ (which is equivalent to measuring distances in units of time), the Lorentz boost equations are as follows (cf. [33]):

$$t = \gamma(t' + ux'), \ x = \gamma(x' + ut'), \ y = y', \ z = z', \tag{3.2}$$

where u, which necessarily has to satisfy $|u| < 1$, is the velocity of the inertial system S' with respect to the inertial system S and $\gamma = (1 - u^2)^{-1/2}$ (the *Lorentz factor*). It follows that the Lorentz transformations, which are composition of spatial rotations and Lorentz boosts, are isometries of (3.1). In fact they are *proper isometries* (their determinant is $\gamma^2(1 - u^2) = +1$) and *orthochronous* (the t and t' variations have the same sign, since $\gamma > 0$).

Conversely, a proper and orthochronous isometry is a Lorentz transformation, since it is easy to see that composed with a suitable rotation is an isometry (still proper and orthochronous) that satisfies $y = y'$, $z = z'$, $t^2 - x^2 = t'^2 - x'^2$, and therefore it is enough to prove, as detailed below, that this transformation is a Lorentz boost.

3.1.1 (Claim) *Let* $t = \delta t' + \delta' x'$, $x = \xi t' + \xi' x'$ *be the equations of a proper orthochronous isometry* f *of* (3.1). *Then* f *is a Lorentz boost.*

Proof The matrix coefficients of f satisfy $\delta\xi' - \delta'\xi = 1$ (for being proper) and $\delta > 0$ (for being orthochronous). In addition, we have the relation

$$t'^2 - x'^2 = (\delta t' + \delta' x')^2 - (\xi t' + \xi' x')^2$$

identically in t' and x' by the condition of isometry. Equating coefficients, we see that this relation is equivalent to the equations

$$\delta^2 - \xi^2 = 1, \quad \delta\delta' - \xi\xi' = 0, \text{ and } \delta'^2 - \xi'^2 = -1.$$

The first equation and the condition $\delta > 0$ allow us to state that there is a unique $\alpha \in \mathbb{R}$ such that $\delta = \cosh\alpha$, $\xi = \sinh\alpha$. From the second equation we infer that there exists $\lambda \in \mathbb{R}$ such that $\delta' = \lambda\xi = \lambda\sinh\alpha$ and $\xi' = \lambda\delta = \lambda\cosh\alpha$. Substituting these values into $\delta\xi' - \delta'\xi = 1$, we get $\lambda = 1$, whereby the third equation is automatically satisfied. Thus the matrix of f has the form $\gamma\begin{pmatrix} 1 & u \\ u & 1 \end{pmatrix}$, with $\gamma = \cosh\alpha$ and $u = \tanh\alpha$. Given that $|u| < 1$ and $\gamma = (1 - u^2)^{-1/2}$, it is clear that f is the Lorentz boost of velocity u (or *rapidity* α). □

In sum, the intrinsic notion corresponding to the group of Lorentz transformations (denoted G_c in [74]) is the group $\mathrm{SO}^+_{1,3}$ of the proper orthochronous isometries of $E_{1,3}$. It is a normal subgroup of the *isometry group* $\mathrm{O}_{1,3}$ of $E_{1,3}$ and of the subgroup $\mathrm{SO}_{1,3} \subset \mathrm{O}_{1,3}$ of the proper isometries.

The essence of *special relativity* is the study of concepts and relations that are invariant by the action of $\mathrm{SO}^+_{1,3}$. It is, therefore, a particular case of Klein's geometry, but its very origin explains its extraordinary potential for expressing statements of geometric and physical content. It is what we try to show in the pages that follow.

Matching Vocabularies: A Mathematical Presentation of Minkowski's Space-Time Geometry

The points of $\mathcal{M}$ will be called *events* and here they will be denoted with capital letters. As in Sect. 1.4, the elements of $E_{1,3}$ will be called *vectors*, and will be denoted with lowercase letters (or capital letters with a dot when they denote derivatives of variable points). This division of roles is necessary, as in Euclidean geometry, to ensure that there are no privileged events (or points).

The *vector separation* between two events P and Q, denoted $Q - P$, is the only vector a such that $Q = P + a$. Recall that in an affine space the sum of points is not defined, and that we have the rules $P + 0 = P$ and $(P + a) + b = P + (a + b)$ for every point P and any vectors a and b.

The *scalar separation* (or, simply, *separation*, or *interval*) between two events P and Q, denoted by $\sigma(P, Q)$, is defined as $\eta(a) = \eta(a, a)$, where $a = Q - P$. As it can be anticipated by what has been said, and as we shall see later, the notion of separation plays a fundamental role in relativistic chronometry and space-time geometry.

To study the properties of separation, it is therefore necessary to study the properties of η. Given $a \in E_{1,3}$, let ϵ_a be the sign of $\eta(a)$ (the *signature* of a). The *magnitude* of a, which will be denoted by $|a|$, is defined as the non-negative real number $|a| = +\sqrt{\epsilon_a \eta(a)}$. This definition, which is equivalent to $\eta(a) = \epsilon_a |a|^2$, is valid for any metric of a real vector space and coincides with the notion of *length* or *norm* of a vector in the case of an Euclidean space E_n. The vectors of magnitude 1 are called *unit vectors*.

For historical reasons (clarified below), positive (negative, null) vectors are also said to be *timelike* (*spacelike*, *lightlike*). And the same terminology applies to the separation as well. We will use both sets of words and let the context suggest which one to choose.

In the following we will assume that $\mathbf{e} = e_0, e_1, e_2, e_3$ is an orthonormal basis with e_0 positive (and therefore with e_1, e_2, e_3 negative). In this chapter, such bases will be called *inertial frames* or simply *frames*. It is important to remark, however, that many authors use the symbol γ_μ instead of e_μ to underline its close relationship to Dirac's Γ_μ matrices (see E.1.8, p. 32), a relation that will be studied in Sect. 3.4.

We will also follow Einstein's *summation criterion* (a repeated index involves a summation with respect to it, unless otherwise indicated) and the convention that the indices designated with Greek letters vary in the set $\{0, 1, 2, 3\}$, whereas those indicated by Latin letters do it in $\{1, 2, 3\}$. For example, if the components of a vector $a \in E_{1,3}$ are denoted a^μ, then $a = a^\mu e_\mu$, whereas $a^k e_k = a - a^0 e_0$. Instead of a^0, it is also customary to use t (*time* coordinate), and x, y, z (*space* coordinates) instead of a_1, a_2, a_3. Thus $\eta(a) = t^2 - (x^2 + y^2 + z^2)$ has the same meaning as

$$\eta(a) = \eta(a^\mu e_\mu) = (a^0)^2 - \left((a^1)^2 + (a^2)^2 + (a^3)^2\right).$$

These expressions of $\eta(a)$ allow us to conclude that $H = \{a \in E_{1,3} : \eta(a) = 1\}$ is a two-sheeted hyperboloid. In terms of the frame $\mathbf{e}$, its sheets are distinguished by the sign of t, but *this sign is not intrinsic*, for the replacement of e_0 by $-e_0$ changes t to $-t$. This indetermination (between two possible indistinguishable temporal orientations) compels us to choose one of the two (let us call it H^+) as the positive *temporal orientation*. This in practice means that only frames $\mathbf{e}$ that satisfy $e_0 \in \mathcal{H}^+$ will be used. We will also assume, to take into account the conclusions at the end of the preceding subsection (p. 55), that any two of these frames (say $\mathbf{e}$ and $\mathbf{e}'$) have the same global orientation. Note that the two assumptions together are equivalent to saying that the isometry determined by $\mathbf{e} \mapsto \mathbf{e}'$ is proper and orthochronous.

As we shall see, H^+ plays a role analogous to that of the sphere S^2 of E_3, and that is why we will call it the *Lorentz sphere*. We will also write

$$F^+ = \mathbb{R}^+ H^+ = \{\lambda u : u \in H^+,\ \lambda \in \mathbb{R}^+\} \tag{3.3}$$

(its elements are the *future-oriented* vectors) and $F^- = -F^+$ (*past-oriented* vectors). The open set $F^+ \cup F^-$ is the *interior* of the *light cone* $\{a \in E : \eta(a) = 0\}$. The exterior of this cone is called *Elsewhere* (see Fig. 3.1a).

The results that follow are the mathematical counterpart used to explain and understand relativistic phenomena that are not very intuitive in the framework of ordinary experience, such as the delay of moving clocks and, in particular, the so-called twin paradox.

3.1.2 (The hyperbolic Cauchy-Schwarz inequality) *If $a, b \in F^+$, then*

$$\eta(a, b) \geqslant |a||b|,$$

and equality occurs if and only if $b = \lambda a$, $\lambda \in \mathbb{R}^+$.

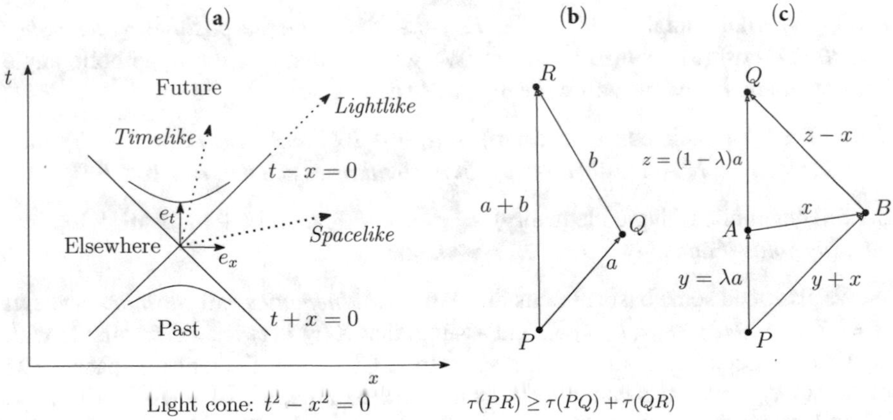

Fig. 3.1 (**a**) Lorentz sphere and types of vectors. (**b**) Hyperbolic triangle inequality (see 3.1.5 and 3.1.6). (**c**) Measuring the spatial distance AB with photons (PB and BQ) and clocks in the timelike segment PQ (see 3.1.8)

Proof There is no loss of generality in assuming that $a, b \in H^+$, in which case we have $\eta(a) = \eta(b) = 1$ and $a^0, b^0 > 0$ (the e_0-components of a and b). Set $\bar{a} = a - a^0 e_0$, $\bar{b} = b - b^0 e_0$, $\alpha = |\bar{a}|$ and $\beta = |\bar{b}|$, so that $-\eta(\bar{a}) = \alpha^2$, $-\eta(\bar{b}) = \beta^2$ and $-\eta(\bar{a}, \bar{b}) \leqslant \alpha\beta$, $\alpha, \beta \geqslant 0$ (we use that $-\eta$ is positive definite in $\langle e_1, e_2, e_3 \rangle$ and so we can apply the Euclidean Cauchy-Schwarz inequality to it). Since

$$1 = \eta(a) = (a^0)^2 + \eta(\bar{a}) = (a^0)^2 - \alpha^2,$$

we have $(a^0)^2 = 1 + \alpha^2$, and similarly $(b^0)^2 = 1 + \beta^2$. It follows that

$$\eta(a, b) = a^0 b^0 + \eta(\bar{a}, \bar{b}) \geqslant a^0 b^0 - \alpha\beta.$$

But

$$(a^0)^2(b^0)^2 = (1+\alpha^2)(1+\beta^2) = 1+\alpha^2+\beta^2+\alpha^2\beta^2 \geqslant 1+2\alpha\beta+\alpha^2\beta^2 = (1+\alpha\beta)^2,$$

which means that $a^0 b^0 \geqslant 1 + \alpha\beta$. So $\eta(a, b) \geqslant 1$.

For the second claim, we may again assume that a and b are unit vectors and what we have to see is that equality holds if and only if $a = b$. In order to achieve equality, the two inequalities used in the proof so far must be an equality. The second is an equality if and only if $\alpha = \beta$, and this gives us that $a_0 = b_0$, inasmuch as

$$a_0^2 = 1 + \alpha^2 = 1 + \beta^2 = b_0^2.$$

On the other hand, the equality $-\eta(\bar{a}, \bar{b}) = \alpha\beta$ is true if and only if $\bar{a} = 0$ or $\bar{b} = 0$ or $\bar{b} = \lambda\bar{a}$, $\lambda > 0$, and it is immediate to check that in all these cases we have $a = b$: if $\bar{a} = 0$, then $\bar{b} = 0$ (for $\beta = \alpha = 0$), and therefore $a = a_0 e_0 = b_0 e_0 = b$; the case $\bar{b} = 0$ is analogous; and in the third case, $\beta = |\bar{b}| = \lambda|\bar{a}| = \lambda\alpha$, from which it follows that $\lambda = 1$ and $a = a_0 e_0 + \bar{a} = b_0 e_0 + \bar{b} = b$. □

3.1.3 (Hyperbolic angle) *If* $a, b \in F^+$, *there is a unique positive real number* δ *such that* $\cosh(\delta) = \eta(a, b)/|a|\,|b|$. We will say that δ is the hyperbolic angle between a and b, and we will write $\delta(a, b)$ to denote it. □

3.1.4 (The hyperbolic cosine theorem) *If* $a, b \in F^+$, *and we set* $\delta = \delta(a, b)$, *then* $(a + b)^2 = a^2 + b^2 + 2|a||b|\cosh(\delta)$. *In particular we see that* $a + b \in F^+$. □

3.1.5 (Hyperbolic triangle inequality) *If* $a, b \in F^+$, *then* $|a + b| \geqslant |a| + |b|$, *and equality holds if and only if* $b = \lambda a$, $\lambda > 0$. □

We also need some basic notions of relativistic *kinematics* and *chronometry*. Let $X = X(s) \in \mathcal{M}$, $s \in [a, b] \subseteq \mathbb{R}$, and assume that $X(s)$ is twice differentiable with continuous second derivative with respect to s. Given that the tangent space $T_X\mathcal{M}$ of $\mathcal{M}$ at any event X is canonically isomorphic to $E_{1,3}$, we have $dX/ds \in E_{1,3}$. We will say that $X(s)$ is a *timelike path* if $dX/ds \in F^+$. Note that this condition is invariant with respect to strictly increasing reparameterizations $s = s(\tau)$, for in this case $dX/d\tau = (dX/ds)(ds/d\tau)$ and $ds/d\tau > 0$. The *proper time* of a timelike path $X(s)$ is the function $\tau : [a, b] \to [0, T]$ defined by

$$\tau(\xi) = \int_0^\xi |dX/ds|\, ds = \int_0^\xi \eta\,(dX/ds)^{1/2}\, ds, \quad T = \tau(b). \tag{3.4}$$

Insofar as $\tau(s)$ is a strictly increasing function of s, we can consider its inverse, $s = s(\tau)$, $\tau \in [0, T]$, and the parameterization $X(\tau) = X(s(\tau))$. Then $dX/d\tau$, which will be denoted $\dot{X}$, satisfies $\dot{X} \in H^+$, and in particular $\eta(\dot{X}) = 1$:

$$\dot{X} = dX/d\tau = (dX/ds)(ds/d\tau) = (dX/ds)/(d\tau/ds) = (dX/ds)/|dX/ds|.$$

Physically, $\tau(s)$ is interpreted as the time ticked by a chronometer that travels along with $X(s)$ and set to 0 at $X(a)$. Since the proper time $T = \tau(b)$ only depends on the curve $\mathcal{X}$ traced by $X(s)$, we can write $\tau(\mathcal{X})$ to denote it.

3.1.6 (The twins' theorem) Let $P, Q \in \mathcal{M}$ and suppose $a = Q - P \in F^+$. Then $X(s) = P + sa$, $s \in [0, 1]$ (geometrically it is the parameterization of the segment PQ joining P and Q) is a timelike path and $\tau(s) = s|a|$, because $dX/ds = a$ for all s and $\eta\,(dX/ds)^{1/2} = |a|$. In particular, $\tau(PQ) = |a|$. Of these kind of timelike paths we say that they are *uniform*. If $b \in F^+$, and we set $R = Q + b = P + (a + b)$, then $a + b \in F^+$ and we have $\tau(PR) \geqslant \tau(PQ) + \tau(QR)$, with equality if and only if $Q \in PR$ (see Fig. 3.1 b). Indeed, $\tau(PQ) = |a|$, $\tau(QR) = |b|$, $\tau(PR) = |a + b|$, and we know that $|a + b| \geqslant |a| + |b|$, with equality if and only if $b = \lambda a$, $\lambda > 0$. □

With the same notations as for the timelike paths, we say that $X(s)$ is a *spacelike path* if dX/ds is spacelike. This condition is also invariant under strictly increasing reparameterizations $s = s(\delta)$ and the *proper distance* of the path $X(s)$ is the function $\delta : [a, b] \to [0, D]$ defined by

$$\delta(\xi) = \int_0^\xi |dX/ds|\, ds = \int_0^\xi (-\eta(dX/ds))^{1/2}\, ds, \quad D = \delta(b). \tag{3.5}$$

Considering that the proper distance $D = \delta(b)$ only depends on the curve $\mathcal{X}$ traced by $X(s)$, we can write $\delta(\mathcal{X})$ to denote it. In the case of a uniform spacelike path (which means that $X(s) = P + sa$, $\eta(a) < 0$, $s \in [0, 1]$, $Q = P + a$), we have $\delta(PQ) = |a|$.

3.1.7 (Meaning of proper time) The proper time of an infinitesimal segment $X(s)X(s + ds)$ of a timelike path $X(s)$ is

$$\tau(X(s)X(s + ds)) = |X(s + ds) - X(s)| = ds\ |dX/ds| = ds\ \eta(dX/ds)^{1/2},$$

which is the integrand of (3.4). In this way we construe proper time as the integral (sum) of uniform infinitesimal proper times. □

A path $X(s)$ is *lightlike* if the vector $dX/ds \in C$ (the light cone). In this case there is no distinguished geometric parameter of the path. Physically, such paths are those traced by *photons*, or zero-mass particles, and they are generatrices of the light cone.

3.1.8 (Distance measurement with clocks and photons; cf. [24, 2.2.1]) *Let $P, Q \in \mathcal{M}$ and suppose $a = Q - P$ is timelike. Let $A = P + \lambda a$, $0 < \lambda < 1$ and $B \in \mathcal{M}$ such that $B - P$ and $Q - B$ are lightlike. Then $x = B - A$ is spacelike and $\delta(AB)^2 = \tau(PA)\tau(AQ)$* (see Fig. 3.1c).

Proof It is clear that $\tau(PA) = \lambda|a|$, $\tau(AQ) = (1-\lambda)|a|$, $\eta(\lambda a + x) = 0$ (as $\lambda a + x = B - P$) and $\eta((1-\lambda)a - x) = 0$ (as $(1-\lambda)a - x = Q - B$). The last two equations give the relations

$$\eta(x) + \lambda^2\eta(a) + 2\lambda\eta(a,x) = 0 \quad \text{and} \quad \eta(x) + (1-\lambda)^2\eta(a) - 2(1-\lambda)\eta(a,x) = 0.$$

Multiplying the first by $1-\lambda$, the second by λ and adding the results, we obtain

$$\eta(x) + \lambda(1-\lambda)\eta(a) = 0.$$

Thus $\eta(x) = -\lambda(1-\lambda)\eta(a) < 0$ (this shows that x is spacelike), and

$$\delta(AB)^2 = \lambda(1-\lambda)\eta(a) = \lambda(1-\lambda)|a|^2 = \tau(PA)\tau(AQ). \qquad \square$$

3.2 A GA View of the Lorentz Group

In Chaps. 1 and 2 we saw how handy the notion of rotor was to construct and operate with isometries, and how flexibly these could be used to deal with geometric transformations, as, for example, the conformal transformations of the Euclidean space E_3. A similar bonus, which includes geometric covariance, also obtains for the spacetime $\mathcal{M}$. In this case the relevant algebra (the Dirac algebra $\mathcal{D}$) was studied in detail in Sect. 1.4 and we will see that the isometries given by rotors in $\mathcal{D}$ provide precise ways to construct and operate with Lorentz transformations.

$\mathcal{D}$-Rotors and Spacetime Isometries

An element $R \in \mathcal{D}^+$ satisfying $R\tilde{R} = 1$ is called a *rotor*.

3.2.1 (Isometry associated with a rotor R) *Let R be a rotor and define $\underline{R} : \mathcal{D} \to \mathcal{D}$ by $\underline{R}(x) = Rx\tilde{R}$. Then:*

(1) *$\underline{R}$ is an automorphism of the geometric product of $\mathcal{D}$.*
(2) $\underline{R}(\mathcal{D}^1) = \mathcal{D}^1$.

(3) *The linear map* $\underline{R} : \mathcal{D}^1 \to \mathcal{D}^1$ *is an isometry.*
(4) $\underline{R}$ *preserves grades and is an automorphism of the exterior and inner products.*

Proof (1) $\underline{R}$ is linear and $\underline{R}(xy) = Rxy\tilde{R} = Rx\tilde{R}Ry\tilde{R} = \underline{R}(x)\underline{R}(y)$ for all $x, y \in \mathcal{D}$. This shows that $\underline{R}$ is an automorphism of the geometric product.

(2) To see that $\underline{R}(a) \in \mathcal{D}^1$ when $a \in \mathcal{D}^1$, note that R and a are even and odd, respectively, and hence $\widehat{\underline{R}(a)} = \hat{R}\hat{a}\hat{\tilde{R}} = -Ra\tilde{R} = -\underline{R}(a)$, which shows that $\underline{R}(a)$ only can have odd grades (1 or 3). But we also have $\widetilde{\underline{R}(a)} = \tilde{\tilde{R}}\tilde{a}\tilde{R} = Ra\tilde{R} = \underline{R}(a)$, which shows that $\underline{R}(a)$ has no grade 3 component. This proves that $\underline{R}(a) \in \mathcal{D}^1$.

(3) $(\underline{R}a)^2 = \underline{R}(a^2) = a^2$ shows that $\underline{R}$ is an isometry. In the first step we have used (1) and in the second that a^2 is a scalar.

(4) The first part follows from the fact that $\underline{R}$ maps a product of k pair-wise orthogonal vectors to a product of k pair-wise orthogonal vectors (by (1) and (3)). From the definition of the exterior product by means of the geometric product, it is immediate to derive that $\underline{R}(\boldsymbol{x}_1 \wedge \cdots \wedge \boldsymbol{x}_k) = \underline{R}\boldsymbol{x}_1 \wedge \cdots \wedge \underline{R}\boldsymbol{x}_k$, and this implies that we also have $\underline{R}(x \wedge y) = \underline{R}x \wedge \underline{R}y$ for all $x, y \in \mathcal{D}$. Finally $\underline{R}(x \cdot y) = \underline{R}x \cdot \underline{R}y$ follows easily from the definition of the inner product by means of the geometric product. □

Although it is feasible to describe all rotors (as, for example, in [97]), for our present purposes it will suffice to construct what we will call *Lorentz rotors*.

Given a bivector $z = \boldsymbol{x} + \boldsymbol{y}\mathbf{i} \in \mathcal{D}^2$, we have $z^2 = \boldsymbol{x}^2 - \boldsymbol{y}^2 + 2(\boldsymbol{x} \cdot \boldsymbol{y})\mathbf{i} \in \mathbf{C}$ (since $\mathbf{i}$ commutes with bivectors, $\boldsymbol{x}\boldsymbol{y}\mathbf{i} + \boldsymbol{y}\mathbf{i}\boldsymbol{x} = 2(\boldsymbol{x} \cdot \boldsymbol{y})\mathbf{i}$, where the interior product is relative to $\mathcal{E}$, and $\boldsymbol{x}^2, \boldsymbol{y}^2, \boldsymbol{x} \cdot \boldsymbol{y} \in \mathbb{R}$). We see that z^2 is real if and only if $\boldsymbol{x} \cdot \boldsymbol{y} = 0$, and in this case we will say that z is a *Lorentz bivector*. If in addition $z^2 = \pm 1 = \epsilon$, we will say that z is a *unit Lorentz bivector*, *positive* or *negative* according to the sign ϵ.

For example, if $\boldsymbol{v} \in \mathcal{E}$ is a unit vector, then $\boldsymbol{v}$ and $\boldsymbol{v}\mathbf{i}$ are unit Lorentz bivectors. The first is positive and the second is negative.

3.2.2 (The rotor $R_{z,\alpha}$) *If z is a unit Lorentz bivector and $\alpha \in \mathbb{R}$, $R_{z,\alpha} = e^{\alpha z/2}$ is a rotor* (the denominator 2 in the exponent is included in order that the significant parameter be α, not 2α)*. Moreover,*

$$R_{z,\alpha} = \cos_\epsilon(\alpha/2) + z\sin_\epsilon(\alpha/2),$$

where $\cos_\epsilon$ *and* $\sin_\epsilon$ *denote* cosh *and* sinh *if z is positive and* cos *and* sin *if z is negative.*

Proof Let $R = R_{z,\alpha}$. Given that $\tilde{z} = -z$, $\tilde{R} = e^{-\alpha z/2}$ and so $R\tilde{R} = 1$. Now in the development of the exponential $e^{\alpha z/2}$, all terms have positive sign if $\epsilon = 1$. The terms with even exponent have the form $\frac{1}{(2k)!}(\alpha/2)^{2k}$, and the terms with odd exponent have the form $z\frac{1}{(2k+1)!}(\alpha/2)^{2k+1}$, and hence it is clear that $R = \cosh(\alpha/2) + z\sinh(\alpha/2)$. The case $\epsilon = -1$ can be established with a similar reasoning. □

3.2.3 (Rotors of Lorentz boosts) *Let* $\boldsymbol{v} \in \mathcal{E}$ *be a unit relative vector and set* $v = \boldsymbol{v}e_0 \in \langle e_1, e_2, e_3 \rangle$ (note that $\boldsymbol{v}$ is the relative vector of v, as $v \wedge e_0 = ve_0 = \boldsymbol{v}$). *Then* $\underline{R}_{\boldsymbol{v},\alpha}$ *is the Lorentz boost in the direction* v *of* rapidity α (and velocity $u = \tanh \alpha$).

Proof Using that e_0 anticommutes with $\boldsymbol{v}$, we have

$$\underline{R}(e_0) = R^2_{\boldsymbol{v},\alpha} e_0 = e^{\alpha \boldsymbol{v}} e_0 = \cosh(\alpha) e_0 + \sinh(\alpha) v,$$

$$\underline{R}(v) = e^{\alpha \boldsymbol{v}/2} v e^{-\alpha \boldsymbol{v}/2} = e^{\alpha \boldsymbol{v}/2} \boldsymbol{v} e^{\alpha \boldsymbol{v}/2} e_0 = e^{\alpha \boldsymbol{v}} v = \sinh(\alpha) e_0 + \cosh(\alpha) v.$$

On the other hand, the vectors $a \in \langle e_0, v \rangle^{\perp}$ are fixed by $\underline{R}$, for they commute with $\boldsymbol{v}$, and the matrix of the restriction of $\underline{R}_{\boldsymbol{v},\alpha}$ to $\langle e_0, v \rangle$ is $\gamma \begin{pmatrix} 1 & u \\ u & 1 \end{pmatrix}$, with $\gamma = \cosh \alpha$ and $u = \tanh \alpha$. Given that $|u| < 1$ and $\gamma = (1 - u^2)^{-1/2}$, it is clear that $\underline{R}_{\boldsymbol{v},\alpha}$ is the Lorentz boost of velocity u in the direction v (cf. the proof of 3.1.1). □

3.2.4 (Relativistic composition of velocities) *The velocity of the composition of two Lorentz boosts of velocities* u_1 *and* u_2 *in the same direction is a boost in that direction of velocity* $u = (u_1 + u_2)/(1 + u_1 u_2)$.

Proof If $\alpha_i \in \mathbb{R}$ are the rapidities of the boosts and v their common direction, then their rotors are $R_i = e^{\alpha_i \boldsymbol{v}/2}$, and $u_i = \tanh(\alpha_i)$ $(i = 1, 2)$. It follows that

$$R_2 R_1 = e^{(\alpha_1 + \alpha_2)\boldsymbol{v}/2}$$

is the rotor of the composition $\underline{R}_2\, \underline{R}_1$, which shows that this composition is a boost in the direction v of velocity

$$\begin{aligned} u &= \tanh(\alpha_1 + \alpha_2) \\ &= (\tanh \alpha_1 + \tanh \alpha_2)/(1 + \tanh \alpha_1 \tanh \alpha_2) \\ &= (u_1 + u_2)/(1 + u_1 u_2). \end{aligned}$$

□

3.2.5 (Rotations) Let $\boldsymbol{v} \in \mathcal{E}$ be a unit relative vector and set $z = \boldsymbol{v}\mathbf{i}$. As observed before, z is a unit negative Lorentz bivector. Let $R = R_{z,\alpha}$. In this case e_0 commutes with z and therefore

$$\underline{R}(e_0) = e^{\alpha z/2} e_0 e^{-\alpha z/2} = e^{\alpha z/2} e^{-\alpha z/2} e_0 = e_0.$$

Thus we see that $\underline{R}$ induces a rotation in $e_0^{\perp} = \langle e_1, e_2, e_3 \rangle$. The axis of this rotation is $v = \boldsymbol{v}e_0$, for v also commutes with z. Finally, the angle of rotation is α, for if $x \in \langle e_1, e_2, e_3 \rangle$ is orthogonal to v, then x anticommutes with z and

$$\underline{R}(x) = e^{\alpha z} x = \cos(\alpha) x + z x \sin(\alpha).$$

Note that $zx = \boldsymbol{v}ix \in \langle e_1, e_2, e_3\rangle$, as it is linear combination of x and $\underline{R}x$, and that it is orthogonal to v and x, as it anticommutes with both. □

We end with a fact that will be useful in the next two sections.

3.2.6

(1) *If* $e_0, e_0' \in H^+$ (unit future pointing timelike vectors), *there is a Lorentz boost* L *such that* $L(e_0) = e_0'$. *The rapidity of this boost is the hyperbolic angle between* e_0 *and* e_0'.
(2) *Any proper orthochronous isometry of* $E_{1,3}$ *is the composition of a rotation and Lorentz boost. In particular, it can be obtained as* $\underline{R}$ *for some rotor* R,

Proof

(1) We may assume e_0 and e_0' are linearly independent (by 3.1.2 this is equivalent to say that $e_0 \neq e_0'$). We set α to denote the hyperbolic angle between e_0 and e_0', so that $e_0 \cdot e_0' = \cosh\alpha$. Pick a unit vector $v \in \langle e_1, e_2, e_3\rangle \cap \langle e_0, e_0'\rangle$ (the latter intersection has dimension 1 and it is enough to normalize any non-zero vector in it). Then we have $\langle e_0, e_0'\rangle = \langle e_0, v\rangle$ and $e_0' = \lambda e_0 + \mu v$ ($\lambda, \mu \in \mathbb{R}$). Now $\cosh\alpha = e_0 \cdot e_0' = \lambda$, $1 = e_0'^2 = \lambda^2 - \mu^2$, and so

$$\mu^2 = \cosh^2\alpha - 1 = \sinh^2\alpha.$$

Since v was determined up to sign, we may take $\mu = \sinh\alpha$. Thus $e_0' = \cosh\alpha e_0 + \sinh\alpha v$ and 3.2.3 shows that the Lorentz boost $\underline{R}_{\boldsymbol{v},\alpha}$, $\boldsymbol{v} = v \wedge e_0$, maps e_0 to e_0'.
(2) If $\mathbf{e}'$ is the image of the $\mathbf{e}$ by a proper orthochronous isometry T, then e_0 and e_0' are future pointing unit timelike vectors. By (1) there is a Lorentz boost L such that $L(e_0) = e_0'$. Then $R = L^{-1}T$ leaves e_0 fixed, so that it is a rotation, and therefore $T = LR$. □

3.2.7 (Composition of general boosts) In 3.2.4 we have seen that the composition of two boosts along the same direction is a boost in that direction. In general, the composition of two boosts will be the composition of a rotation and a boost (by the previous result) and the rotation is required when the directions of the boosts are different.

3.3 A GA View of Electrodynamics

First we will see that the GA view of the differential operator d (familiar from the calculus texts) is the *Dirac operator* ∂. This is not an idle translation from one formalism to another, for we soon discover that GA confers many powerful features that cannot be phrased within the ordinary multivariable calculus.

We will need the notion of *reciprocal frame* of a frame e_0, e_1, e_2, e_3. It is defined as the frame e^0, e^1, e^2, e^3 such that $e^0 = e_0$ and $e^k = -e_k$. In general it is determined by the relations $e^\mu \cdot e_\nu = \delta^\mu_\nu$. The components of a vector a with respect to the reciprocal frame are denoted a_μ, so that $a = a_\mu e^\mu$. We clearly have $a_0 = a^0$ and $a_k = -a^k$.

Dirac's Operator

If f is a differentiable function defined in an open set U of $\mathcal{M}$, its differential is given by $df = (\partial_\mu f)\, dx^\mu$, where $\partial_\mu = \partial/\partial x^\mu$. It is a 1-form defined on U, so that we have a linear map $d_x f : E \to \mathbb{R}$ for each $x \in U$. Its value on a vector a is

$$(d_x f)(a) = df(x + at)/dt\,|_{t=0} = (D_a f)(x),$$

where $(D_a f)(x) = df(x + at)/dt|_{t=0}$ is the *directional derivative* of f at x in the direction a. Indeed, $f(x + at) - f(x) = (d_x f)(ta) + o(t) = t(d_x f)(a) + o(t)$, by definition of d_x.

Therefore we can represent the operator d (*differential*) in the form $dx^\mu \partial_\mu$. In this expression, dx^μ is the linear map $E \to \mathbb{R}$ such that $(dx^\mu)(e_\nu) = \delta^\mu_\nu$. Since this form coincides with $e^\mu\cdot$, we see that d is realized by the operator $\partial = e^\mu \partial_\mu$. This is the *vector operator* of $\mathcal{D}$, or *Dirac's operator*, and by definition $(\partial f)\cdot a = a\cdot(\partial f)$ is the directional derivative of f in the direction a. In fact this shows that the operator $a \cdot \partial = (a \cdot e^\mu)\partial_\mu = a^\mu \partial_\mu$ yields, when applied to a function f, its derivative in the direction a.

Using ∂ instead of d has other advantages in the context of $\mathcal{D}$. The most relevant is that we can form, for any *multivector field* $F = F(x)$, the products ∂F, $\partial \cdot F$ and $\partial \wedge F$. For example, if $F = F^J e_J$, then

$$\partial \cdot F = e^\mu \partial_\mu F^J \cdot e_J = \partial_\mu F^J\, e^\mu \cdot e_J.$$

In fact, since $e^\mu e_J = e^\mu \cdot e_J + e^\mu \wedge e_J$, we have the relation

$$\partial F = \partial \cdot F + \partial \wedge F$$

for any F.

If $F = F^\nu e_\nu$ is a *vector field*, $\partial \cdot F = \partial_\mu F^\nu e^\mu \cdot e_\nu = \partial_\mu F^\nu \delta^\mu_\nu = \partial_\mu F^\mu$, which is the (Lorentzian) *divergence* of F. In a similar way we see, still in the same case, that $\partial \wedge F = \partial_\mu F^\nu e^\mu \wedge e_\nu$. In the Euclidean space E_3, the vector operator is denoted by ∇, $\nabla \cdot F = \partial_k F^k$ is the usual divergence, $\nabla \wedge F$ is the *curl* of F in bivector form (see E.3.2, p. 72). So it stands to reason to call $\partial \cdot F$ and $\partial \wedge F$ the *divergence* and *curl* of F.

Another important observation is that ∂ provides a solution of Dirac's dream (to find a square root of the *dalembertian* operator $\Box$):

$$\partial^2 = \partial \cdot \partial = \partial_0^2 - (\partial_1^2 + \partial_2^2 + \partial_3^2) = \Box.$$

The Lab Picture

We will use the formalism and results introduced in the last subsection of Sect. 1.4 (page 23) about the relative space $\mathcal{E}$.

The map $E_{1,3} \to \mathcal{E}$, $x \mapsto \boldsymbol{x} = x \wedge e_0$, is surjective and its kernel is $\langle e_0 \rangle$. If $x = x^\mu e_\mu$, it is clear that $\boldsymbol{x} = x^k \boldsymbol{\sigma}_k$. Setting $t = x \cdot e_0$, we have

$$xe_0 = x \cdot e_0 + x \wedge e_0 = t + \boldsymbol{x},$$

which is the representation of x in terms or the relative space $\mathcal{E}$. This *relative representation* is also called *lab representation.*

Taking an event O as origin, the lab representation of an event P is the lab representation of $x = P - O$. For example, the lab representation of $P = O + \tau e_0$ is $t = \tau$ and $\boldsymbol{x} = 0$, which is interpreted as the time given by a clock at rest with respect in $\mathcal{E}$. Another example: the lab representation of the Lorentz quadratic form agrees with the expression used by Minkowski,

$$\eta(x) = x^2 = xe_0e_0x = (t + \boldsymbol{x})(t - \boldsymbol{x}) = t^2 - \boldsymbol{x}^2.$$

The relative expression of the *relativistic velocity* $\dot{x} = dx/d\tau$ is

$$\dot{x}e_0 = d(xe_0)/d\tau = d(t + \boldsymbol{x})/d\tau,$$

which implies

$$dt/d\tau = \dot{x} \cdot e_0, \quad d\boldsymbol{x}/d\tau = \dot{x} \wedge e_0\,.$$

If $\boldsymbol{v} = d\boldsymbol{x}/dt$ (*relative velocity*),

$$\boldsymbol{v} = \frac{d\boldsymbol{x}}{dt} = \frac{d\boldsymbol{x}}{d\tau}\frac{d\tau}{dt} = \frac{\dot{x} \wedge e_0}{\dot{x} \cdot e_0}\,.$$

Given that $\boldsymbol{v}^2 = -\eta(\boldsymbol{v}) = -\eta(\dot{x} \wedge e_0)/(\dot{x} \cdot e_0)^2 = 1 - (\dot{x} \cdot e_0)^{-2}$, we have $\boldsymbol{v}^2 < 1$ and

$$\dot{x} \cdot e_0 = 1/\sqrt{1 - \boldsymbol{v}^2}, \tag{3.6}$$

which is the *Lorentz factor* of $\boldsymbol{v}$ and denoted by $\gamma = \gamma(\boldsymbol{v})$. In particular, $dt = \gamma(\boldsymbol{v})\, d\tau$, or $d\tau = dt\,\sqrt{1 - \boldsymbol{v}^2}$, which gives the precise relation between proper time and the time measured in the lab. Since $\gamma > 1$ if $\boldsymbol{v} \neq 0$, the time measured in the lab is greater than proper time.

The *relativistic moment* of a particle is defined by the formula $p = m_0\dot{x}$, where m_0 is the *rest mass*. Since $\dot{x}e_0 = d(t + \boldsymbol{x})/d\tau = \gamma + \gamma\boldsymbol{v}$,

$$pe_0 = m_0\gamma + m_0\gamma\boldsymbol{v} = m + m\boldsymbol{v} = m + \boldsymbol{p},$$

where $m = \gamma m_0$ is called the *relativistic mass* of the particle and $\boldsymbol{p} = m\boldsymbol{v}$ is moment (relative to the lab). Now it is immediate that

$$\dot{p}e_0 = \gamma\, dm/dt + \gamma\, d\boldsymbol{p}/dt. \tag{3.7}$$

The relative representation of the vector operator ∂ is

$$\partial e_0 = \partial \cdot e_0 + \partial \wedge e_0 = \partial_0 + \boldsymbol{\partial}.$$

In this case, $\boldsymbol{\partial} = \partial \wedge e_0 = e^k \wedge e_0\, \partial_k = -\boldsymbol{\sigma}_k \partial_k = -\boldsymbol{\nabla}$, where $\boldsymbol{\nabla} = \boldsymbol{\sigma}_k \partial_k$ (the vector operator of the relative space). We also have that $e_0\partial = \partial_0 - \boldsymbol{\partial} = \partial_0 + \boldsymbol{\nabla}$.

Riesz Form of Maxwell's Equations

3.3.1 (The Maxwell equation, [82]) *Let* $\boldsymbol{E}, \boldsymbol{B}, \boldsymbol{j} \in \mathcal{E}$ *be time-dependent relative vectors and* $\rho = \rho(\boldsymbol{x}, t)$ *a differentiable function of* $\boldsymbol{x} \in \mathcal{E}$ *and* $t \in \mathbb{R}$. *Define* $F = \boldsymbol{E} + \boldsymbol{B}\mathbf{i}$ (it is called the Faraday bivector) *and* $J = (\rho + \boldsymbol{j})e_0$. *Then the equation*

$$\partial F = J$$

is equivalent to the four Maxwell's equations for the electric field $\mathbf{E}$ *and the magnetic field* $\boldsymbol{B}$ *created by the charge density* ρ *and the current density vector* $\boldsymbol{j}$.

Proof Since $Je_0 = \rho + \boldsymbol{j}$, we have $\rho = J \cdot e_0$ and $\boldsymbol{j} = J \wedge e_0$. Moreover, from

$$(\rho + \boldsymbol{j})e_0 = e_0(\rho - \boldsymbol{j})$$

we obtain that $e_0 J = \rho - \boldsymbol{j}$. Multiplying the equation $\partial F = J$ by e_0 on the left, we obtain the equivalent relation $(\partial_0 + \boldsymbol{\nabla})(\boldsymbol{E} + \mathbf{i}\boldsymbol{B}) = \rho - \boldsymbol{j}$. With a little algebra to develop the products, we get

$$\partial_0 \boldsymbol{E} + \boldsymbol{\nabla} \cdot \boldsymbol{E} + \boldsymbol{\nabla} \wedge \boldsymbol{E} + \mathbf{i}(\partial_0 \boldsymbol{B} + \boldsymbol{\nabla} \cdot \boldsymbol{B} + \boldsymbol{\nabla} \wedge \boldsymbol{B}) = \rho - \boldsymbol{j}.$$

On equating the corresponding grades of both sides, we see that this equation is equivalent to the four equations

$$\boldsymbol{\nabla} \cdot \boldsymbol{E} = \rho, \quad \partial_0 \boldsymbol{E} + \mathbf{i}\boldsymbol{\nabla} \wedge \boldsymbol{B} = -\boldsymbol{j}, \quad \mathbf{i}\partial_0 \boldsymbol{B} + \boldsymbol{\nabla} \wedge \boldsymbol{E} = 0, \quad \mathbf{i}\boldsymbol{\nabla} \cdot \boldsymbol{B} = 0.$$

Now it suffices to observe that $\boldsymbol{\nabla}\cdot$ is the divergence operator of the relative space and that $\mathbf{i}\boldsymbol{\nabla} \wedge \boldsymbol{B} = -\boldsymbol{\nabla} \times \boldsymbol{B} = -\operatorname{curl}(\boldsymbol{B})$ (the *curl vector* of $\boldsymbol{B}$) to conclude that these equations are equivalent to

$$\operatorname{div}(\boldsymbol{E}) = \rho \qquad \text{(Gauss law for } \boldsymbol{E}) \tag{3.8}$$

$$\text{curl}(\boldsymbol{B}) - \partial_t \boldsymbol{E} = \boldsymbol{j} \qquad \text{(Ampère-Maxwell law)} \tag{3.9}$$

$$\partial_t \boldsymbol{B} + \text{curl}(\boldsymbol{E}) = 0 \qquad \text{(Faraday's induction law)} \tag{3.10}$$

$$\text{div}(\boldsymbol{B}) = 0 \qquad \text{(Gauss law for } \boldsymbol{B}) \tag{3.11}$$

which are Maxwell's equations in differential form for the electromagnetic field created by ρ and $\boldsymbol{j}$ (in units such that $\epsilon_0 = \mu_0 = 1$, and hence also $c = 1$). □

3.3.2 (Invariants) *The expression F^2 is clearly Lorentz invariant. In terms of the decomposition $F = \boldsymbol{E} + \boldsymbol{B}\mathbf{i}$, $F^2 = \boldsymbol{E}^2 - \boldsymbol{B}^2 + 2(\boldsymbol{E} \cdot \boldsymbol{B})\mathbf{i}$, and so $\boldsymbol{E}^2 - \boldsymbol{B}^2$ and $\boldsymbol{E} \cdot \boldsymbol{B}$ are Lorentz invariant.* □

3.3.3 (The continuity equation) If we multiply $\partial F = J$ by ∂ on the left, we obtain $\square F = \partial \cdot J + \partial \wedge J$. Since the left side is a bivector ($\square$ preserves grades), the scalar part of the right-hand side expression must vanish: $\partial \cdot J = 0$. This is the *charge conservation equation,* as it is equivalent, in relative terms, to the *continuity equation* $\partial_t \rho + \nabla \cdot \boldsymbol{j} = 0$. □

The Relativistic Lorentz Force Law

Consider a particle with electric charge q in an electromagnetic field F. We will see that the relativistic form of the *Lorentz force law* is the *Einstein-Lorentz* relation

$$\dot{p} = qF \cdot \dot{x}, \tag{3.12}$$

where $\dot{x}$ is the proper velocity and p the relativistic moment of the particle.

3.3.4 (Lab form of the Lorentz force law) *If $F = \boldsymbol{E} + \boldsymbol{i}\boldsymbol{B}$ is the expression of F in relative terms, then the Einstein-Lorentz formula is equivalent to the relations*

$$dm/dt = q(\boldsymbol{E} \cdot \boldsymbol{v}) \ \ \text{and} \ \ d\boldsymbol{p}/dt = q(\boldsymbol{E} + \boldsymbol{B} \times \boldsymbol{v}). \tag{3.13}$$

Proof The relative expression of the vector $qF \cdot \dot{x}$ is $q(F \cdot \dot{x}) \cdot e_0 + q(F \cdot \dot{x}) \wedge e_0$. The scalar part is $q(F \cdot \dot{x}) \cdot e_0 = qF \cdot (\dot{x} \wedge e_0) = \gamma q F \cdot \boldsymbol{v} = \gamma q \boldsymbol{E} \cdot \boldsymbol{v}$, because $(\boldsymbol{i}\boldsymbol{B}) \cdot \boldsymbol{v} = 0$ (the inner products are in $\mathcal{D}$, not in $\mathcal{P}$; see E.3.4 and E.3.5, p. 73). This and the formula (3.7) show that the first formula of our statement is equivalent to the equality of the e_0-components of the Einstein-Lorentz relation. Let us now turn to $q(F \cdot \dot{x}) \wedge e_0 = q(\boldsymbol{E} \cdot \dot{x}) \wedge e_0 + q(\boldsymbol{i}\boldsymbol{B} \cdot \dot{x}) \wedge e_0$. The first summand is equal to $\gamma q \boldsymbol{E}$, because

$$(\boldsymbol{\sigma}_k \cdot \dot{x}) \wedge e_0 = (e_k e_0 \cdot \dot{x}) \wedge e_0 = (\dot{x} \cdot e_0) e_k e_0 = \gamma \boldsymbol{\sigma}_k.$$

And the second summand is equal to $\gamma q \boldsymbol{B} \times \boldsymbol{v}$, because a similar calculation gives (with jkl a cyclic permutation of 123)

$$(i\boldsymbol{\sigma}_j \cdot \dot{x}) \wedge e_0 = (-(e_k e_l) \cdot \dot{x}) \wedge e_0 = \dot{x}_k \boldsymbol{\sigma}_l - \dot{x}_l \boldsymbol{\sigma}_k$$
$$= \gamma(\boldsymbol{\sigma}_l v_k - \boldsymbol{\sigma}_k v_l) = \gamma \boldsymbol{\sigma}_j \times \boldsymbol{v},$$

where we have set $\boldsymbol{v} = v_k \boldsymbol{\sigma}_k$ and $\dot{x} = \dot{x}_\mu e_\mu$. With the last two relations, and the formula (3.7), the proof is complete. □

3.3.5 ($m \sim E$) In the second formula (3.13), the term $\boldsymbol{f} = d\boldsymbol{p}/dt$ is the force exerted by the electromagnetic field on the particle and the term on its right is the *Lorentz law* for this force. The power of the Lorentz force is $\boldsymbol{f} \cdot \boldsymbol{v} = q\boldsymbol{E} \cdot \boldsymbol{v}$, because $\boldsymbol{B} \times \boldsymbol{v}$ is perpendicular to $\boldsymbol{v}$. Now the first of the formulas (3.13) tells us that this power is equal to dm/dt, which means that the variations of relativistic mass are equivalent to energy. In fact, the work produced on the particle by the Lorentz force in a time interval, which is to say the integral of the power on that interval, is equal to the variation of mass in that interval. We may conclude that $p_0 = p \cdot e_0 = m$ has the form $m = w + w_0$, where w denotes the dynamical energy of the particle and w_0 a constant that only depends on the rest mass m_0.

Since m_0 is itself a mass, it can be equated to an energy, and in fact we may think that it is the energy needed for its creation, or the energy liberated in its disintegration, so that finally it makes sense to write $m = E$, where E is the sum of the dynamical energy w and the energy that corresponds to m_0. In SI units, the formula looks more familiar: $E = mc^2$.

Potentials

Equating the grade components of the two sides of $\partial F = J$, we see that the equality is equivalent to the equations $\partial \cdot F = J$ and $\partial \wedge F = 0$ (which correspond to the first and second pairs of Maxwell's equations, namely the non-homogeneous equations pair, (3.8) and (3.9), and the homogeneous equations pair, (3.10) and (3.11), respectively).

The second equation (together with the Poincaré lemma) tells us that there is a vector field A (which is called a *potential* of F) such that $F = \partial \wedge A$. In this form the equation $\partial \wedge F = 0$ is automatically satisfied, for $\partial \wedge \partial = 0$, and the equation $\partial \cdot F = J$ becomes $\partial \cdot (\partial \wedge A) = J$, that is, $(\partial \cdot \partial)A - \partial(\partial \cdot A) = J$, or $\Box A - \partial(\partial \cdot A) = J$. Let us see that we can choose A so that it satisfies $\partial \cdot A = 0$ (*Lorenz gauge condition*, or *Lorenz gauge*, after Ludvig Lorenz). Indeed, if f is a scalar function, then $\partial \wedge \partial f = 0$ and so $\partial \wedge (A + \partial f) = \partial \wedge A = F$, and the point is that we can find f so that $\partial \cdot (A + \partial f) = 0$, as this condition is equivalent to the equation $\Box f = -\partial \cdot A$ (for the existence of a solution of this equation, see, for example, [53]). With the Lorentz condition, we have $F = \partial A$, and the equation $\partial \cdot F = J$ becomes $\Box A = J$. This is the (non-homogeneous) *wave equation* for A. If we manage to solve this equation, given J and appropriate boundary conditions, then we can get F by computing ∂A.

In relative terms, $Ae_0 = \phi + \boldsymbol{A}$, with $\phi = A \cdot e_0$ and $\boldsymbol{A} = A \wedge e_0 \in \mathcal{E}$. Then we have $\partial A = \partial e_0 e_0 A = (\partial_t - \nabla)(\phi - \boldsymbol{A}) = -(\nabla\phi + \partial_t \boldsymbol{A}) + \nabla \wedge \boldsymbol{A}$, for

$\partial_t \phi + \nabla \cdot \boldsymbol{A} = \partial \cdot A = 0$ (Lorentz condition). Equating corresponding grades with $\boldsymbol{E} + \boldsymbol{B}\boldsymbol{i}$, we see that the relation $\partial A = F$ is equivalent to the equations

$$\boldsymbol{E} = -(\nabla\phi + \partial_t \boldsymbol{A}), \quad \boldsymbol{B} = -\boldsymbol{i}(\nabla \wedge \boldsymbol{A}) = \nabla \times \boldsymbol{A} = \mathrm{curl}(\boldsymbol{A}), \tag{3.14}$$

which are the familiar formulas that supply the electric and magnetic fields in the lab from ϕ (*scalar potential*) and $\boldsymbol{A}$ (*vector potential*).

Relative Transformation of the Electromagnetic Field

One of the more emblematic results of Einstein's special theory of relativity [33] is the relation between the electric and magnetic fields observed in two inertial frames. The obtention of such relations can be posed as follows. We know that the transformation that maps $\mathbf{e}$ into another inertial frame $\mathbf{e}'$ is proper and orthochronous. Thus there is a rotor R (see 3.2.6) such that

$$e'_\mu = \underline{R}(e_\mu) = R e_\mu \tilde{R}.$$

Now the fundamental observation is the following (note the use of geometric covariance for $\underline{R}$):

$$E'_k = \boldsymbol{\sigma}'_k \cdot F = \underline{R}(\boldsymbol{\sigma}_k) \cdot F = \boldsymbol{\sigma}_k \cdot \underline{\tilde{R}}(F) = \boldsymbol{\sigma}_k \cdot \tilde{R} F R,$$
$$B'_k = \boldsymbol{\sigma}'^*_k \cdot F = \underline{R}(\boldsymbol{\sigma}^*_k) \cdot F = \boldsymbol{\sigma}^*_k \cdot \tilde{R} F R.$$

We see that the problem is reduced to the calculation of $\tilde{R} F R$.

Let us work out in detail the case in which R is a Lorentz boost, say (with notations from Example 3.2.3),

$$R = R_{\boldsymbol{\sigma}_1,\alpha} = e^{\alpha\boldsymbol{\sigma}_1/2}.$$

We recall that $\underline{R}$ is the Lorentz boost in the direction $e_1 = \boldsymbol{\sigma}_1 e_0$ whose velocity is $u = \tanh\alpha$ and $\gamma = \cosh\alpha$.

Relatively to $\mathbf{e}$, we can write

$$F = E_1\boldsymbol{\sigma}_1 + E_2\boldsymbol{\sigma}_2 + E_3\boldsymbol{\sigma}_3 + B_1\boldsymbol{\sigma}^*_1 + B_2\boldsymbol{\sigma}^*_2 + B_3\boldsymbol{\sigma}^*_3.$$

Since R commutes with $\boldsymbol{\sigma}_1$ and anticommutes with $\boldsymbol{\sigma}_2$ and $\boldsymbol{\sigma}_3$,

$$\tilde{R} F R = E_1\boldsymbol{\sigma}_1 + e^{-\alpha\boldsymbol{\sigma}_1}(E_2\boldsymbol{\sigma}_2 + E_3\boldsymbol{\sigma}_3) + B_1\boldsymbol{\sigma}^*_1 + e^{-\alpha\boldsymbol{\sigma}_1}(B_2\boldsymbol{\sigma}^*_2 + B_3\boldsymbol{\sigma}^*_3).$$

To transform this expression we have that

$$e^{-\alpha\boldsymbol{\sigma}_1} = \cosh\alpha - \boldsymbol{\sigma}_1 \sinh\alpha = \gamma(1 - u\boldsymbol{\sigma}_1)$$

and the relations

$$\sigma_1\sigma_2 = \sigma_3^*, \quad \sigma_1\sigma_3 = -\sigma_2^*, \quad \sigma_1\sigma_2^* = -\sigma_3, \quad \sigma_1\sigma_3^* = \sigma_2.$$

With all that we get

$$\begin{aligned}\tilde{R}FR &= E_1\sigma_1 + \gamma(E_2\sigma_2 + E_3\sigma_3) - \gamma u(E_2\sigma_3^* - E_3\sigma_2^*) \\ &+ B_1\sigma_1^* + \gamma(B_2\sigma_2^* + B_3\sigma_3^*) + \gamma u(B_2\sigma_3 - B_3\sigma_2).\end{aligned}$$

Setting $\boldsymbol{E}_\parallel = E_1\boldsymbol{\sigma}_1$ and $\boldsymbol{E}_\perp = E_2\boldsymbol{\sigma}_2 + E_3\boldsymbol{\sigma}_3$, with analogous notations for $\boldsymbol{B}$, and taking into account that $E_2\sigma_3 - E_3\sigma_2 = \boldsymbol{\sigma}_1 \times \boldsymbol{E}$, and similarly for $\boldsymbol{B}$, one obtains the following version of *Einstein's formulas* (writing $\boldsymbol{u} = u\boldsymbol{\sigma}_1$):

$$\begin{aligned}\boldsymbol{E}' &= \boldsymbol{E}_\parallel + \gamma\boldsymbol{E}_\perp + \boldsymbol{u} \times \boldsymbol{B}, \\ \boldsymbol{B}' &= \boldsymbol{B}_\parallel + \gamma\boldsymbol{B}_\perp - \boldsymbol{u} \times \boldsymbol{E}.\end{aligned}$$

3.3.6 (Remark) These formulas show that both $\boldsymbol{E}$ and $\boldsymbol{B}$ are involved in the calculation of $\boldsymbol{E}'$, and the same is true for $\boldsymbol{B}'$. A specially compelling example is a particle at rest at the origin of $\mathbf{e}'$, whose field is reduced to an electric Coulomb field $\boldsymbol{E}'$. Then Einstein's second formula gives us (inverting the roles of $\mathbf{e}$ and $\mathbf{e}'$)

$$\boldsymbol{B} = \boldsymbol{u} \times \boldsymbol{E}'.$$

This shows that the reference $\mathbf{e}$ measures a magnetic field in addition to the electric field $\boldsymbol{E} = \boldsymbol{E}'_\parallel + \gamma\boldsymbol{E}'_\perp$. The conclusion is important: *the magnetic fields created by moving charges are relativistic effects of the Coulomb fields created by stationary charges.* □

3.3.7 (Remark) Einstein's formulas are valid for any Lorentz boost. This is seen by replacing the $\boldsymbol{\sigma}_1$ in the rotor by the unit vector $\boldsymbol{u}$ giving the direction of the boost and defining the parallel and perpendicular components of $\boldsymbol{E}$ and $\boldsymbol{B}$ with respect to it. In this case the boost velocity is $u\boldsymbol{u}$ and the proof is similar, but using the rotor $R_{\boldsymbol{u},\alpha}$ instead of $R_{\boldsymbol{\sigma}_1,\alpha}$. □

3.4 A GA View of Dirac's Equation

In Schrödinger's theory, the values of the *wave function* $\psi(x)$, $x \in \mathcal{M}$, are complex numbers. When Pauli introduced the spin, he was lead to replace $\mathbb{C}$ by $\mathbb{C}^2$, and this is why since then the (normalized) vectors of $\mathbb{C}^2$ are called *Pauli spinors*. Then Dirac was lead to replace $\mathbb{C}^2$ by $\mathbb{C}^4$ (space of *Dirac spinors*, or also *bispinors*) because his matrices Γ_μ were 4×4 complex matrices. The landmarks of the road followed by Dirac can be summarized as follows.

As starting point, take the *Klein-Gordon equation*, $(\Box + m^2)\psi = 0$. This equation is the Schrödinger equation for an electron, with $m = m_e/\hbar$, m_e the rest-mass of the electron and $\hbar = h/(2\pi)$ the Planck constant (normalized). It is a partial differential equation of the second order, but Dirac argued that what was needed was a linear equation in ∂_t and, requiring that it be relativistic, also linear in ∂_x, ∂_y and ∂_z. So he considered the operator $D = d_\mu \partial_\mu$, and on imposing that $D^2 = \Box$ he discovered that the simplest solution was to take $d_\mu = \Gamma_\mu$ (the matrices introduced in E.1.8, p. 32).

Now the Klein-Gordon equation factors as $(D - im)(D + im)\psi = 0$, and Dirac simply postulated the equation $(D + im)\psi = 0$, which is equivalent to the form in which he wrote it: $i\hbar D\psi = m_e\psi$. But in order that this equation makes sense, ψ has to take values in $\mathbb{C}^4$, and with this assumption it is the *Dirac equation*. In the presence of an electromagnetic potential A, it is included in the Dirac equation in the form $i\hbar(D - eA)\psi = m_e\psi$, where e is the electron charge, but note that it is also required that A has the form $A_\mu \Gamma_\mu$.

The problem of expressing the Dirac equation purely in terms of the algebra $\mathcal{D}$ was studied for the first time by Hestenes [40], a theme on which he has provided many decisive insights over the years, as in [41–46], culminating so far in the masterpieces [47] (for perspective on his work, see also the New Preface to the second edition of [40] and the paper [48]). We have found that all these works are particularly instructive, and also others, as, for example, the pedagogical [13] or the treatise [29]. The effort is worthwhile, since on the one hand it is possible to dispense with burdens that are accidental to the problem (some have been mentioned before), and on the other, the rich structure of $\mathcal{D}$ can be exploited to delve further into the understanding of phenomena.

The Hestenes formulation of Dirac's equation (cf. [41, §2]) reads:

$$\partial\psi\, i\,\hbar - eA\psi = m_e\psi e_0. \tag{3.15}$$

The nature of its ingredients is as follows. The field ψ (the *Dirac field*) takes values in the even algebra $\mathcal{D}^+$ ($\psi : \mathcal{M} \to \mathcal{D}^+$). Note that the complex dimension of $\mathcal{D}^+$ is 4, the same as $\mathbb{C}^4$. The expression $\partial\psi$ is the geometric product of the Dirac operator ∂ with ψ, while $A\psi$ is the geometric product of the vector potential A and ψ. The symbol i is the geometric imaginary unit $i = e_2e_1 = \mathbf{i}e_3e_0 = \mathbf{i}\sigma_3 = \sigma_1\sigma_2$. Therefore we see that in the Equation (3.15), which Hestenes calls *real Dirac equation*, and which here we will dub the *Hestenes-Dirac equation*, all terms are geometrically meaningful. No matrices appear in it, and even the formal imaginary unit i has been replaced by the geometric area unit $\sigma_1\sigma_2$.

But the real value gained with (3.15) is the possibility of exploiting the rich structure of $\mathcal{D}^+$ to sharpen our understanding of the electron theory. A first fundamental result is the following [41]:

3.4.1 (Canonical form of ψ) *If $\psi \in \mathcal{D}^+$ and $\psi\tilde{\psi} \neq 0$, there exist $\rho \in \mathbb{R}^+$, $\beta \in (-\pi, \pi]$ and a rotor R such that $\psi = \rho^{1/2} e^{\beta\,\mathbf{i}/2} R$. Moreover, this expression is unique.*

Proof The product $\psi\tilde{\psi}$ is a complex scalar (its grade 2 term must vanish because it is invariant by reversion). Therefore we can write it in polar form: $\psi\tilde{\psi} = \rho e^{\beta\,\mathbf{i}}$, with $\rho > 0$ and $\beta \in (-\pi, \pi]$. Let $R = \rho^{-1/2}\psi e^{-\beta\,\mathbf{i}/2}$. Then R is a rotor, because $\tilde{R} = \rho^{-1/2} e^{-\beta\,\mathbf{i}/2}\tilde{\psi}$ and $R\tilde{R} = \rho^{-1}\psi e^{-\beta\,\mathbf{i}}\tilde{\psi} = 1$. Uniqueness is also clear, for ρ and β are uniquely determined by $\psi\tilde{\psi}$, and R is uniquely determined by the relation $R = \psi\rho^{-1/2} e^{-\beta\,\mathbf{i}/2}$. □

The role of R in the study of the Dirac-Hestenes equation is that $\underline{R}$ is a proper and orthochronous isometry. That it is proper follows from the fact that R commutes with $\mathbf{i}$, so that $\underline{R}\mathbf{i} = R\mathbf{i}\tilde{R} = \mathbf{i}R\tilde{R} = \mathbf{i}$. On the other hand, $\underline{R}$ is orthochronous because it is the composition of a rotation and a Lorentz boost.

The rotor R is in fact a *rotor field* and so it enables us to construct the frame field $\mathbf{e}' = \underline{R}\mathbf{e}$, all with the same orientation and temporal orientation as $\mathbf{e}$.

3.4.2 (Frame field of a ψ and Dirac's current) *We have the relation $\psi e_\mu \tilde{\psi} = \rho e'_\mu$. Setting $v = e'_0$, we have $\psi e_0 \tilde{\psi} = \rho v$* (which is the *Dirac current*).

Proof Indeed, given that $\mathbf{i}$ anticommutes with vectors and that $\tilde{\mathbf{i}} = \mathbf{i}$, we can write:

$$\psi e_\mu \tilde{\psi} = \rho e^{\beta\,\mathbf{i}/2} R e_\mu \tilde{R} e^{\beta\,\mathbf{i}/2} = \rho e^{\beta\,\mathbf{i}/2} e^{-\beta\,\mathbf{i}/2} R e_\mu \tilde{R} = \rho e'_\mu.$$ □

The vector $s = \frac{\hbar}{2} R e_3 \tilde{R} = \frac{\hbar}{2} e'_3$ is the *spin vector*. The rotor R transforms the geometric bivector unit $i = e_2 e_1$ into $\iota = R i \tilde{R} = e'_2 e'_1$ and $S = \frac{\hbar}{2}\iota$ is the *spin bivector*.

3.4.3 (Meaning of the spin bivector) $S = \mathbf{i} s v$

Proof $\mathbf{i} s v = \frac{\hbar}{2}\mathbf{i} R e_3 \tilde{R} R e_0 \tilde{R} = \frac{\hbar}{2} R \mathbf{i} e_3 e_0 \tilde{R} = \frac{\hbar}{2} R e_2 e_1 \tilde{R} = \frac{\hbar}{2} R i \tilde{R} = \frac{\hbar}{2}\iota = S.$ □

3.5 Exercises

E.3.1 Write a proof of 3.1.5.

E.3.2 (The **div** and **curl** Operators of E_3) If $F = F^k \boldsymbol{e}_k$ is a vector field on E_3, its divergence, $\mathbf{div}(F)$, is defined as $\partial_1 F^1 + \partial_2 F^2 + \partial_3 F^3$. Check that this agrees with $\nabla \cdot F$ (recall that ∇ denotes the vector operator of E_3). On the other hand, the curl of F, denoted $\mathbf{curl}(F)$, is defined as $\sum(\partial_j F^k - \partial_k F^j)\boldsymbol{e}_l$, where jkl runs over the cyclic permutations of 123. Show that $\nabla \wedge F = \sum(\partial_j F^k - \partial_k F^j)\boldsymbol{e}_j \wedge \boldsymbol{e}_k$ (same summation convention). Since $\boldsymbol{e}_j \wedge \boldsymbol{e}_k = -i\boldsymbol{e}_l$, we see that $\nabla \wedge F$ is the Hodge dual of $\mathbf{curl}(F)$. *Hint:* In E_3, any orthonormal basis is its own reciprocal basis.

E.3.3 If $F \in \mathcal{D}^2$, and we split it as $F = \boldsymbol{E} + \boldsymbol{B}\mathbf{i}$ in the Pauli algebra, show that then $F' = \boldsymbol{E} - \boldsymbol{B}\mathbf{i}$ is the reversal of F in that algebra and that

$$FF' = \boldsymbol{E}^2 + \boldsymbol{B}^2 + 2\boldsymbol{E} \times \boldsymbol{B}.$$

As shown in the standard treatises on electromagnetism, $\boldsymbol{E}^2 + \boldsymbol{B}^2$ measures the density of the electromagnetic field energy, while $\boldsymbol{E} \times \boldsymbol{B}$, which is called the *Poynting vector*, controls the flow of energy across surfaces.

E.3.4 Prove that if $a, b \in \mathcal{D}^1$ and $F \in \mathcal{D}^k$, $k \geqslant 2$, then $a \cdot (b \cdot F) = (a \wedge b) \cdot F$. Use the commutation rule for the inner product to see that the latter relation is equivalent to $(F \cdot a) \cdot b = F \cdot (a \wedge b)$, which is the relation used at the beginning of the proof of (3.13) in the case $k = 2$.

E.3.5 Prove the relation $\mathbf{i}\boldsymbol{B} \cdot \boldsymbol{v} = 0$ used in the proof of (3.13), where the inner product is in $\mathcal{D}$.

E.3.6 (Monochromatic electromagnetic waves in vacuum) An electromagnetic field F of the form $F = F_0 e^{\mathbf{i}(k \cdot x)}$ (where $F_0 \in \mathcal{D}^2$ is a constant non-zero bivector and $k \in \mathcal{D}^1$ a constant non-zero vector) is said to be a *monochromatic wave*. In vacuum (absence of charges and currents), the Riesz-Maxwell equation for these waves is $\partial F = 0$, and the steps below sketch how to derive well-known properties of them.

(1) Show that $\partial(k \cdot x) = k$ and $\partial F = k\mathbf{i}F_0 e^{\mathbf{i}(k \cdot x)}$. So $\partial F = 0$ is equivalent to $kF_0 = 0$.
(2) Writing $ke_0 = \omega + \boldsymbol{k}$ in the lab formalism, and hence $e_0 k = \omega - \boldsymbol{k}$, we have $k^2 = ke_0\epsilon_0 k = \omega^2 - \boldsymbol{k}^2$. Therefore, multiplying $kF_0 = 0$ by k, we get $\omega^2 = \boldsymbol{k}^2$ (this is called the *dispersion relation* and its meaning will be seen in a moment).
(3) Let $F_0 = \boldsymbol{E} + \mathbf{i}\boldsymbol{B}$, with $\boldsymbol{E}, \boldsymbol{B} \in \mathcal{E}$. Then we have $(\omega - \boldsymbol{k})(\boldsymbol{E} + \mathbf{i}\boldsymbol{B}) = 0$, which is equivalent to $\omega\boldsymbol{E} + \mathbf{i}\omega\boldsymbol{B} = \boldsymbol{k} \cdot \boldsymbol{E} + \boldsymbol{k} \wedge \boldsymbol{E} + \mathbf{i}(\boldsymbol{k} \cdot \boldsymbol{B}) + \mathbf{i}(\boldsymbol{k} \wedge \boldsymbol{B})$. Looking at the scalar part, we get $\boldsymbol{k} \cdot \boldsymbol{E}$. Similarly, the pseudoscalar part yields $\boldsymbol{k} \cdot \boldsymbol{B}$.
(4) So we have $\omega\boldsymbol{E} + \mathbf{i}\omega\boldsymbol{B} = \boldsymbol{k} \wedge \boldsymbol{E} + \mathbf{i}(\boldsymbol{k} \wedge \boldsymbol{B})$, and from this it is easy to conclude that $\omega\boldsymbol{E} = -\boldsymbol{k} \times \boldsymbol{B}$ and $\omega\boldsymbol{B} = \boldsymbol{k} \times \boldsymbol{E}$. This shows that $\boldsymbol{k}, \boldsymbol{E}, \boldsymbol{B}$ is an orthogonal system, and also that $\boldsymbol{E}^2 = \boldsymbol{B}^2$.

Chapter 4
Robot Kinematics

In this chapter we present a general overview of the kinematics of serial robotic manipulators based on the conformal geometric algebra, which simplifies its formulation and provides means for solving, clearly and efficiently, several of the main classical problems in this field. The core of the material presented in this chapter is assembled from Sect. 4.2 onwards, and for convenience of the reader Sect. 4.1 is devoted to provide a brief review of the classical approach to robot kinematics.

4.1 Classical Kinematics

A *serial robot manipulator* is an open kinematic chain made up of a sequence of rigid bodies, called *links*, connected by means of actuated kinematic pairs, called *joints*, that provide relative motion between consecutive links. At the end of the last link, there is a tool or device known as the *end-effector*. Only two types of joints are considered throughout this chapter: *revolute joints*, that only perform rotations, and *prismatic joints*, that only perform translations.

From a kinematic point of view, the end-effector position and orientation (*pose*) of a manipulator can be expressed as a differentiable function $f : C \to X$, where C denotes the *space of joint variables*, called *configuration space*, and X denotes the end-effector configuration space, which is usually called the *operational space*.

For serial manipulators, a frame $\{\boldsymbol{o}_i, \boldsymbol{x}_i, \boldsymbol{y}_i, \boldsymbol{z}_i\}$ is attached to each joint of the manipulator to describe its relative position and orientation (Fig. 4.1a). The relations between consecutive joint frames are described using the *Denavit-Hartenberg convention* [26]. This convention consists of four parameters, the D-H parameters (Fig. 4.1b): one acting as a *joint variable*, either an angle θ_i or a displacement d_i, depending on whether the joint i is revolute or prismatic; and the other three acting as constants: length a_i, angle α_i and either d_i or θ_i depending on which

C. Lavor et al., *A Geometric Algebra Invitation to Space-Time Physics, Robotics and Molecular Geometry*, SpringerBriefs in Mathematics,
https://doi.org/10.1007/978-3-319-90665-2_4

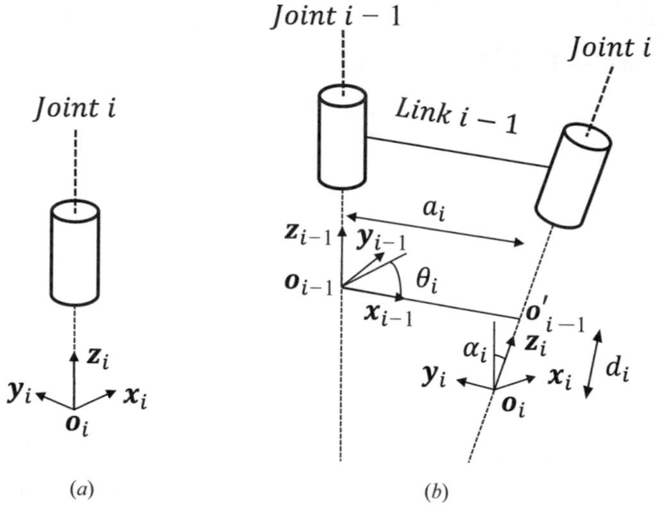

Fig. 4.1 (**a**) Frame attached to joint i. (**b**) Relation between consecutive joints given by the four D-H parameters: a_i, the perpendicular distance between the joint axes z_{i-1} and z_i; α_i, the angle between the joint axes z_{i-1} and z_i; θ_i, the angle between $\boldsymbol{x}_{i-1}$ and $\boldsymbol{x}_i$; and d_i, the distance between $\boldsymbol{o}'_{i-1}$ and $\boldsymbol{o}_i$

one is used to describe the joint variable [88, 90]. Therefore, associated with each joint i, together with the corresponding orthonormal frame (Fig. 4.2), there is a transformation matrix ${}^{i-1}T_i$ that relates frame $\{i\}$ to the preceding one (the first joint frame is related to the world frame). The end-effector pose ${}^0T_n$ can be represented as follows:

$$ {}^0T_n = {}^0T_1 \, {}^1T_2 \, \cdots \, {}^{n-1}T_n \tag{4.1}$$

with

$$ {}^0T_n = \begin{pmatrix} R & \boldsymbol{p} \\ 0 & 1 \end{pmatrix}, \tag{4.2}$$

where R is a rotation matrix that describes the end-effector orientation with respect to the world frame, while $\boldsymbol{p}$ is a position vector describing the end-effector position with respect to the world frame.

This description is equivalent to the one provided by f, known as the *kinematic function* of the serial robot. Thus, $f(\boldsymbol{q}) = \boldsymbol{x}$, where $\boldsymbol{x}$ denotes the vector describing the end-effector pose, and $\boldsymbol{q} = (q_1, \ldots, q_n)$, the vector whose components are the joint variables, also known as the *configuration*. Clearly, either $q_i = \theta_i$ if joint i is revolute or $q_i = d_i$ if joint i is prismatic. Deriving with respect to time the kinematic relation provided by f, we obtain another relation

$$ \dot{\boldsymbol{x}} = J(\boldsymbol{q})\dot{\boldsymbol{q}}, \tag{4.3}$$

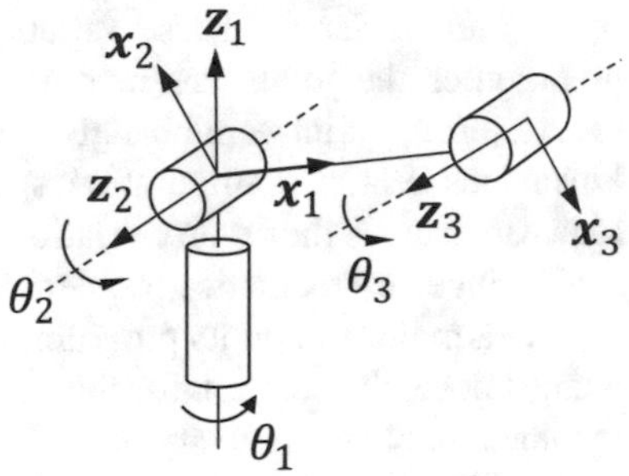

Fig. 4.2 Schematic representation of a spatial serial robot manipulator of three DoF

where $\dot{\boldsymbol{x}}$ denotes the end-effector velocity vector; $\dot{\boldsymbol{q}}$, the vector of the joint velocities; and J, the *Jacobian matrix* of f. If $f(\boldsymbol{q}) = (f_1(\boldsymbol{q}), \ldots, f_6(\boldsymbol{q}))$, then $J = (\partial f_i(\boldsymbol{q})/\partial q_j)_{ij}$. In robot kinematics, this Jacobian is also known as the *analytical Jacobian matrix* and is denoted by $J_A(\boldsymbol{q})$.

However, $\dot{\boldsymbol{x}}$ does not correspond to the vector of linear and angular velocities of the end-effector. Therefore, the following relation is established:

$$J_A(\boldsymbol{q}) = T(\boldsymbol{x}) J_G(\boldsymbol{q}), \tag{4.4}$$

where $T(\boldsymbol{x})$ denotes the transformation matrix that is only singular at *representation singularities* [88], i.e., a singularity associated with an adopted minimal representation of the orientation and $J_G(\boldsymbol{q})$, the *geometric Jacobian matrix* of the serial robot.

4.1.1 *If* $J_G = (J_1, \ldots, J_n)$, *then:*

$$\begin{aligned} J_i &= \begin{bmatrix} z_i \times (\boldsymbol{o}_n - \boldsymbol{o}_i) \\ z_i \end{bmatrix} && \text{if } i \text{ is revolute,} \\ J_i &= \begin{bmatrix} z_i \\ 0 \end{bmatrix} && \text{if } i \text{ is prismatic.} \end{aligned} \tag{4.5}$$

In robotics, the geometric Jacobian J_G is the most used because, computationally speaking, it is easier to calculate and more efficient if implemented in different algorithms. Furthermore, the geometric Jacobian relates the vector of joint velocities $\dot{\boldsymbol{q}}$ with the vector of linear and angular velocities of the end-effector $\boldsymbol{v}$:

$$\boldsymbol{v} = J_G(\boldsymbol{q})\dot{\boldsymbol{q}}. \tag{4.6}$$

A manipulator is said to have *n degrees of freedom* (DoF) if its configuration can be minimally specified by n variables. For a serial robot, the number and nature of the joints determine the number of DoF. For the task of positioning and orientating its end-effector in the three-dimensional space, the manipulators with more than 6 DoF are called *redundant* while the rest are *non-redundant*.

Robot kinematics includes the study of the *forward* and *inverse kinematics*. Forward kinematics consists of obtaining the pose of the end-effector given the value of the joint variables. These variables can be angles or displacements, depending on whether the joints are revolute or prismatic. Inverse kinematics consists of recovering the joint variables given the end-effector pose. In other words, forward kinematics relates configuration space C to operational space X, while inverse kinematics gives the reverse relation.

The inverse kinematics problem is of special importance for serial robots since it plays a major role in its programming, commanding and control. However, given a target pose, the solution for this problem may not be unique, since non-redundant manipulators have up to sixteen different solution configurations for the same pose [81], while for redundant manipulators this number is infinite. For example, if a trajectory in Cartesian space is considered, each point of such trajectory represents a target pose of the end-effector. Thus, the high number of solutions allows the selection of those with, for example, lower energy consumption or lower joint velocities (these are examples of *secondary tasks*). Therefore, one of the main objectives when solving the inverse kinematics is to obtain all the solutions for a given end-effector pose. However, this still is an open problem of great relevance.

The methods used for solving the inverse kinematics are categorized into two groups:

(a) Analytical or closed-form methods: All the solutions are expressed as functions in terms of the pose elements.
(b) Numerical methods: Starting with an initial configuration $\boldsymbol{q}_0$, an iterative process returns a good approximation $\widetilde{\boldsymbol{q}}$ of one of the solutions.

Closed-form methods strongly depend on the geometry of the manipulator and, therefore, are not sufficiently general, i.e., they cannot be applied to arbitrary robots. However, it is clear that they have advantages over the numerical methods such as lower computational cost and execution time. Besides, they give all the solutions for a given end-effector pose. In his PhD thesis, Pieper [81] develops a procedure for obtaining the solutions for a class of serial manipulators, i.e., the manipulators with three consecutive joints whose axes are either parallel or intersect at a single point (if these joints are the last three, these manipulators are known as *manipulators with a spherical wrist*). Later, Paul [77] establishes a more rigorous and generic method based on the manipulation of the transformation matrices that can be applied to arbitrary manipulators. Other methods include the use of Lagrange multipliers, the definition of imaginary links for redundant manipulators, the definition of an extra angle (the arm angle parameter), and different geometric methods [16, 51, 52, 71, 98].

On the other hand, *numerical methods* usually work for any manipulator, but they suffer from several drawbacks like high computational cost and execution time, existence of local minima and numerical errors. Moreover, only one of the sixteen (infinite) possible solutions is obtained for non-redundant (redundant) manipulators. The most well-known numerical approaches are the Jacobian-based methods, in which the relation (4.6) is inverted and solved iteratively. Inverting the geometric Jacobian matrix is not always possible. For redundant manipulators, $J_G(\boldsymbol{q})$ is a non-square matrix, while for non-redundant robots $\det(J_G(\boldsymbol{q}))$ vanishes at singularities.

To handle these situations, alternative methods like the pseudoinverse, the transpose, the damped least-squares and local optimization are used [10, 11, 25]. Other numerical methods include the use of an augmented Jacobian matrix and Crank-Nicholson methods [32, 34, 91].

As mentioned before, singularities do not allow the implementation of simple schemes for solving the inverse kinematics. Besides, singularities affect the motion of the robot. More precisely, a *singularity* is a configuration in which the robot loses some of its degrees of freedom and, hence, motion in the operational space X.

Using the relation (4.6), it is easy to see that:

4.1.2 *If* $\boldsymbol{q} \in C$ *is a singularity of a given serial robot, then the following two statements hold:*

- *The end-effector cannot be translated or rotated around at least one Cartesian direction.*
- *Finite linear and angular velocities of the end-effector may require infinite joint velocities.*

Any serial manipulator of $n > 2$ DoF has singularities as demonstrated in [36, 50]. The identification of such singularities is made by solving the following non-linear equation:

$$\det(J_G(\boldsymbol{q})) = 0 \tag{4.7}$$

if the robot is non-redundant, and

$$\det(J_G(\boldsymbol{q})J_G^T(\boldsymbol{q})) = 0 \tag{4.8}$$

if it is redundant.

In general, if the serial robot possesses at least one revolute joint, several coefficients of the geometric Jacobian matrix are non-linear expressions and, thus, neither Eq. (4.7) nor Eq. (4.8) is easy to formulate and solve. However, for manipulators with a spherical wrist, a simplification can be made. Since the origin of the frame attached to the end-effector can be placed at the common intersection point, also known as the *wrist center point*, a zero block appears in $J_G(\boldsymbol{q})$ by definition (see Eq. (4.5)). Hence,

$$J_G(\boldsymbol{q}) = \begin{pmatrix} J_{11}(\boldsymbol{q}) & 0 \\ J_{21}(\boldsymbol{q}) & J_{22}(\boldsymbol{q}) \end{pmatrix}, \tag{4.9}$$

where $J_{11}(\boldsymbol{q})$, $J_{21}(\boldsymbol{q})$ are blocks of order $3 \times (n-3)$ and $J_{22}(\boldsymbol{q})$ is a block of order 3. Clearly, Eq (4.7) is simplified to:

$$\det(J_G(\boldsymbol{q})) = \det(J_{11}(\boldsymbol{q}))\det(J_{22}(\boldsymbol{q})), \tag{4.10}$$

from where the singularities can be obtained as solutions of either $\det(J_{11}(\boldsymbol{q})) = 0$ or $\det(J_{22}(\boldsymbol{q})) = 0$. These two equations allow to decouple the singularities into position and orientation singularities as follows:

- Position singularities $PS = \{\boldsymbol{q} \in C : \det(J_{11}(\boldsymbol{q})) = 0\}$;
- Orientation singularities $OS = \{\boldsymbol{q} \in C : \det(J_{22}(\boldsymbol{q})) = 0\}$.

4.2 Forward Kinematics

The conformal model of the three-dimensional geometric algebra $\mathcal{C}$ provides an elegant and compact way of describing the forward kinematics of serial robots. While the classical approach is based on the transformation matrices constructed following the D-H convention, this approach needs only the elements of the rotor group.

Let us consider the screw motion related to an arbitrary axis z given by the translation and rotation around such axis. Both motions can be represented using rotors and, therefore, the screw motion can also be represented by a rotor. To establish such representation, let us consider the basis elements of the world frame $\{\boldsymbol{e}_1, \boldsymbol{e}_2, \boldsymbol{e}_3\}$, i.e., the canonical basis of the Euclidean space E_3. Since the i-th joint frame is always described with respect to the $(i-1)$-th joint frame, if the latter is considered as the world frame, the i-th joint frame is described with respect to $\{\boldsymbol{e}_1, \boldsymbol{e}_2, \boldsymbol{e}_3\}$.

Now, the following rotors are defined:

$$
\begin{aligned}
\boldsymbol{T}_{d_i} &= 1 + \frac{d_i \boldsymbol{e}_\infty \boldsymbol{e}_3}{2}, \\
R_{\theta_i} &= \cos\left(\frac{\theta_i}{2}\right) - \sin\left(\frac{\theta_i}{2}\right) \boldsymbol{e}_{12}, \\
\boldsymbol{T}_{a_i} &= 1 + \frac{a_i \boldsymbol{e}_\infty \boldsymbol{e}_1}{2}, \\
R_{\alpha_i} &= \cos\left(\frac{\alpha_i}{2}\right) - \sin\left(\frac{\alpha_i}{2}\right) \boldsymbol{e}_{23},
\end{aligned}
\tag{4.11}
$$

where $\boldsymbol{e}_{ij} = \boldsymbol{e}_i \wedge \boldsymbol{e}_j$. Then, the rotors describing the screw motion are:

$$
\begin{aligned}
M_{\theta_i} &= T_{d_i} R_{\theta_i}, \\
M_{\alpha_i} &= T_{a_i} R_{\alpha_i}.
\end{aligned}
\tag{4.12}
$$

Notice that rotor M_{θ_i} contains both joint variables, while M_{α_i} is a constant rotor. Some authors [5, 6] denote these rotors as *motors* due to their connections with the screw theory (a motor is seen as a motion combining a translation and a rotation around the same axis).

Given a geometric entity (point, line, plane, etc.) x, the following relation is applied in the same way as a general rotor:

$$x' = M_{\theta_i} x \widetilde{M}_{\theta_i} = \boldsymbol{T}_{d_i} R_{\theta_i} x \widetilde{R}_{\theta_i} \widetilde{\boldsymbol{T}}_{d_i}, \tag{4.13}$$

where x' denotes the geometric entity x once it has been rotated and translated.

Analogously to the classical approach, where the product of the transformation matrices defines the forward kinematics (as shown in Eq. (4.1)), the successive multiplication of rotors given by

$$\begin{aligned} x' &= M_{\theta_1} M_{\alpha_1} \cdots M_{\theta_n} M_{\alpha_n} x \widetilde{M}_{\alpha_n} \widetilde{M}_{\theta_n} \cdots \widetilde{M}_{\alpha_1} \widetilde{M}_{\theta_1} \\ &= M_1 \cdots M_n x \widetilde{M}_n \cdots \widetilde{M}_1, \end{aligned} \tag{4.14}$$

with $M_i = M_{\theta_i} M_{\alpha_i}$, also determines the forward kinematics of serial robots. This statement can be easily proven: Eq. (4.14) is valid for points (that represent the end-effector position) and lines (that represent the end-effector orientation).

4.2.1 *Let us consider a particular configuration $\boldsymbol{q} = (q_1, \ldots, q_n) \in C$. Then, the end-effector pose P' associated with $\boldsymbol{q}$ is the multivector:*

$$P' = M_1(q_1) \cdots M_n(q_n) P \widetilde{M}_n(q_n) \cdots \widetilde{M}_1(q_1), \tag{4.15}$$

where P denotes the initial pose of the end-effector.

As mentioned before, the advantages of this approach include a compact representation of the forward kinematics. In addition, since geometric entities and rotors are elements of the algebra, the manipulation of complex geometric structures (like serial chains) becomes easier. Moreover, at computational level, it has lower execution time and computational cost. By using conformal geometric algebra, the matrix products become multivector products, avoiding a significant number of operations.

To illustrate these advantages, a simple example is introduced. Let us consider a planar manipulator of three links (Fig. 4.3a). According to the D-H parameters depicted in Fig. 4.3b, rotor M_i is given by

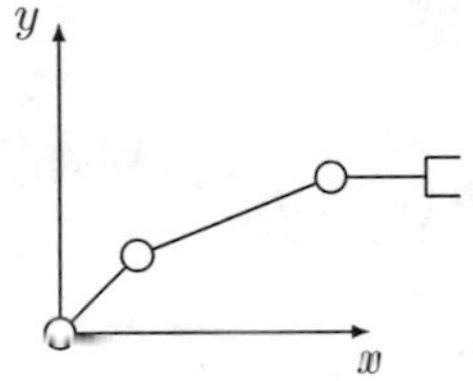

(a)

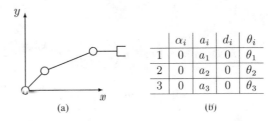

	α_i	a_i	d_i	θ_i
1	0	a_1	0	θ_1
2	0	a_2	0	θ_2
3	0	a_3	0	θ_3

(b)

Fig. 4.3 Planar manipulator: (**a**) Schematic representation. (**b**) D-H parameters

$$
\begin{aligned}
M_i &= M_{\theta_i} M_{\alpha_i} = \overbrace{\boldsymbol{T}_{d_i}}^{1} R_{\theta_i} \boldsymbol{T}_{a_i} \overbrace{R_{\alpha_i}}^{1} \\
&= \left(\cos\left(\frac{\theta_i}{2}\right) - \sin\left(\frac{\theta_i}{2}\right)\boldsymbol{e}_{12}\right)\left(1 + \frac{a_i \boldsymbol{e}_\infty \boldsymbol{e}_1}{2}\right) \\
&= \cos\left(\frac{\theta_i}{2}\right) - \frac{\cos(\theta_i/2)a_i}{2}\boldsymbol{e}_1\boldsymbol{e}_\infty - \sin\left(\frac{\theta_i}{2}\right)\boldsymbol{e}_{12} - \frac{\sin(\theta_i/2)a_i}{2}\boldsymbol{e}_2\boldsymbol{e}_\infty,
\end{aligned} \tag{4.16}
$$

where $\cos(\theta_i/2)$ is the scalar part and the remaining terms conform a grade-two element. Let us denote by B_i this grade two element. Then, since the planar manipulator has 3 DoF, the product of the three motors M_1, M_2, and M_3 can be written as:

$$
\begin{aligned}
M_1 M_2 M_3 &= \left(\cos\left(\frac{\theta_1}{2}\right) - B_1\right)\left(\cos\left(\frac{\theta_2}{2}\right) - B_2\right)\left(\cos\left(\frac{\theta_3}{2}\right) - B_3\right) \\
&= \cos\left(\frac{\theta_1}{2}\right)\cos\left(\frac{\theta_2}{2}\right)\cos\left(\frac{\theta_3}{2}\right) - \cos\left(\frac{\theta_1}{2}\right)\cos\left(\frac{\theta_3}{2}\right)B_2 \\
&\quad - \cos\left(\frac{\theta_2}{2}\right)\cos\left(\frac{\theta_3}{2}\right)B_1 - \cos\left(\frac{\theta_1}{2}\right)\cos\left(\frac{\theta_2}{2}\right)B_3 \\
&\quad + \cos\left(\frac{\theta_1}{2}\right)B_2 B_3 + \cos\left(\frac{\theta_2}{2}\right)B_1 B_3 + \cos\left(\frac{\theta_3}{2}\right)B_2 B_3 \\
&\quad - B_1 B_2 B_3,
\end{aligned} \tag{4.17}
$$

where

$$
\begin{aligned}
B_i B_j = &\left(\frac{\cos(\theta_i/2)\sin(\theta_j/2)a_i - \cos(\theta_j/2)\sin(\theta_i/2)a_j}{2}\right)\boldsymbol{e}_2\boldsymbol{e}_\infty \\
&+ \left(\frac{\sin(\theta_i/2)\sin(\theta_j/2)a_j - \sin(\theta_i/2)\sin(\theta_j/2)a_i}{2}\right)\boldsymbol{e}_1\boldsymbol{e}_\infty \\
&- \sin\left(\frac{\theta_i}{2}\right)\sin\left(\frac{\theta_j}{2}\right)
\end{aligned} \tag{4.18}
$$

and

$$
\begin{aligned}
B_1 B_2 B_3 = &-\left(C_1 \sin\left(\frac{\theta_3}{2}\right) + C_3\frac{\cos(\theta_3/2)a_3}{2}\right)\boldsymbol{e}_1\boldsymbol{e}_\infty \\
&+ \left(C_2 \sin\left(\frac{\theta_3}{2}\right) - C_3\frac{\sin(\theta_3/2)a_3}{2}\right)\boldsymbol{e}_2\boldsymbol{e}_\infty - C_3 \sin\left(\frac{\theta_3}{2}\right)\boldsymbol{e}_{12},
\end{aligned} \tag{4.19}
$$

with

$$C_1 = \frac{\cos(\theta_1/2)\sin(\theta_2/2)a_1 - \cos(\theta_2/2)\sin(\theta_1/2)a_2}{2},$$
$$C_2 = \frac{\sin(\theta_1/2)\sin(\theta_2/2)a_2 - \sin(\theta_1/2)\sin(\theta_2/2)a_1}{2},$$
$$C_3 = \sin\left(\frac{\theta_1}{2}\right)\sin\left(\frac{\theta_2}{2}\right).$$

For each configuration $\boldsymbol{q} = (\theta_1, \theta_2, \theta_3)$, the elements C_1, C_2, and C_3 are real constant numbers. Therefore, the forward kinematics of the planar manipulator is determined by Eq. (4.17):

$$P' = M_1(\theta_1)M_2(\theta_2)M_3(\theta_3)P\widetilde{M}_3(\theta_3)\widetilde{M}_2(\theta_2)\widetilde{M}_1(\theta_1). \tag{4.20}$$

4.3 Differential Kinematics

Differential kinematics refers to the relation (4.6), that allows to obtain the vector $\boldsymbol{v}$ of linear and angular velocities of the end-effector given the joint velocities $\dot{\boldsymbol{q}}$.

In robot kinematics, Eq. (4.6) is used in the design of robust control algorithms for commanding a serial robot in the execution of complex tasks. As in the preceding case (the forward kinematics), conformal geometric algebra provides a framework that avoids the use of matrices. In this context, the algorithms designed will exhibit a better performance with less execution time.

In this section, an analogous of Eq. (4.6) is obtained in terms of rotors. This development roughly follows the one introduced in [7, 99]. It starts by differentiating Eq. (4.15), which yields

$$\dot{P}' = \sum_{j=1}^{n} \frac{\partial}{\partial q_j}\left(\prod_{i=1}^{n} M_i P \prod_{i=1}^{n} \widetilde{M}_{n-i+1}\right)\dot{q}_j, \tag{4.21}$$

where the dependence on the configuration is omitted for simplicity. For the case $n = 2$, (4.21) becomes

$$\begin{aligned}\dot{P}' &= \left(\frac{\partial}{\partial q_1}M_1M_2P\widetilde{M}_2\widetilde{M}_1 + M_1M_2P\widetilde{M}_2\frac{\partial}{\partial q_1}\widetilde{M}_1\right)\dot{q}_1 \\ &+ \left(M_1\frac{\partial}{\partial q_2}M_2P\widetilde{M}_2\widetilde{M}_1 + M_1M_2P\widetilde{M}_2\frac{\partial}{\partial q_1}\widetilde{M}_1\right)\dot{q}_2.\end{aligned} \tag{4.22}$$

Clearly, Eq. (4.22) can be extended and regrouped for an arbitrary n as follows:

$$\dot{P}' = \sum_{j=1}^{n} \left[\frac{\partial}{\partial q_j} \left(\prod_{i=1}^{j} M_i \right) A_1 + A_2 \frac{\partial}{\partial q_j} \left(\prod_{i=n-j+1}^{n} \widetilde{M}_{n-i+1} \right) \right] \dot{q}_j, \tag{4.23}$$

where

$$\begin{aligned} A_1 &= \prod_{i=j+1}^{n} M_i P \prod_{i=1}^{n} \widetilde{M}_{n-i+1}, \\ A_2 &= \prod_{i=1}^{n} M_i P \prod_{i=1}^{n-j} \widetilde{M}_{n-i+1}. \end{aligned} \tag{4.24}$$

As stated in Exercise E.4.3, the derivative of a rotor R with respect to time is

$$\dot{R} = -\frac{1}{2} B R \quad \text{and} \quad \dot{\widetilde{R}} = \frac{1}{2} \widetilde{R} B, \tag{4.25}$$

where B is the bivector associated with R. Thus, the derivatives of the products of rotors can be rewritten as

$$\begin{aligned} \frac{\partial}{\partial q_j} \left(\prod_{i=1}^{j} M_i \right) &= -\frac{1}{2} \left(\prod_{i=1}^{j-1} M_i \right) B_j M_j, \\ \frac{\partial}{\partial q_j} \left(\prod_{i=n-j+1}^{n} \widetilde{M}_{n-i+1} \right) &= \frac{1}{2} \widetilde{M}_j B_j \prod_{i=n-j+2}^{n} \widetilde{M}_{n-i+1}, \end{aligned} \tag{4.26}$$

where B_j denotes the bivector of rotor M_j.

By substitution of (4.26) in (4.23), the following expression is obtained:

$$\dot{P}' = \sum_{j=1}^{n} \left[-\frac{1}{2} \left(\prod_{i=1}^{j-1} M_i \right) B_j M_j A_1 + \frac{1}{2} A_2 \widetilde{M}_j B_j \prod_{i=n-j+2}^{n} \widetilde{M}_{n-i+1} \right] \dot{q}_j. \tag{4.27}$$

Now, (4.27) can be simplified by re-expressing A_1 and A_2

$$\dot{P}' = \sum_{j=1}^{n} \left[-\frac{1}{2} \left(\prod_{i=1}^{j-1} M_i \right) B_j A_1 + \frac{1}{2} A_2 B_j \prod_{i=n-j+2}^{n} \widetilde{M}_{n-i+1} \right] \dot{q}_j, \tag{4.28}$$

where $A_1 = \prod_{i=j}^{n} M_i P \prod_{i=1}^{n} \widetilde{M}_{n-i+1}$ and $A_2 = \prod_{i=1}^{n} M_i P \prod_{i=1}^{n-j+1} \widetilde{M}_{n-i+1}$.

Using the property

$$\prod_{i=1}^{j-1} M_i \prod_{i=1}^{j-1} \widetilde{M}_{j-i} = 1 \quad \forall j = 1, \dots, n, \tag{4.29}$$

identity (4.28) becomes

$$\begin{aligned} \dot{P}' = \sum_{j=1}^{n} \Bigg[& -\frac{1}{2} \prod_{i=1}^{j-1} M_i B_j \prod_{i=1}^{j-1} \widetilde{M}_{j-i} \prod_{i=1}^{j-1} M_i A_1 \\ & + \frac{1}{2} A_2 \prod_{i=1}^{j-1} \widetilde{M}_{j-i} \prod_{i=1}^{j-1} M_i B_j \prod_{i=1}^{j-1} \widetilde{M}_{j-i} \Bigg] \dot{q}_j. \end{aligned} \tag{4.30}$$

Clearly, since the conformal geometric algebra version of the forward kinematics relation (4.14) is valid for any multivector, the following identity is deduced:

$$B_j' = \prod_{i=1}^{j-1} M_i B_j \prod_{i=1}^{j-1} \widetilde{M}_{j-i}, \tag{4.31}$$

where B_j and B_j' are bivectors. These bivectors can be regarded as defining the rotation plane normal to the axis of rotor M_j. Therefore, Eq. (4.30) remains as

$$\dot{P}' = \sum_{j=1}^{n} \left[-\frac{1}{2} B_j' \prod_{i=1}^{j-1} M_i A_1 + \frac{1}{2} A_2 \prod_{i=1}^{j-1} \widetilde{M}_{j-i} B_j' \right] \dot{q}_j. \tag{4.32}$$

Finally, since

$$\begin{aligned} \prod_{i=1}^{j-1} M_i A_1 &= \prod_{i=1}^{j-1} M_i \prod_{i=j}^{n} M_i P \prod_{i=1}^{n} \widetilde{M}_{n-i+1} \\ &= \prod_{i=1}^{n} M_i P \prod_{i=1}^{n} \widetilde{M}_{n-i+1} = P' \end{aligned} \tag{4.33}$$

and

$$\begin{aligned} A_2 \prod_{i=1}^{j-1} \widetilde{M}_{j-i} &= \prod_{i=1}^{n} M_i P \prod_{i=1}^{n-j+1} \widetilde{M}_{n-i+1} \prod_{i=1}^{j-1} \widetilde{M}_{j-i} \\ &= \prod_{i=1}^{n} M_i P \prod_{i=1}^{n} \widetilde{M}_{n-i+1} = P', \end{aligned} \tag{4.34}$$

Eq. (4.32) is simplified to

$$\dot{P}' = \sum_{j=1}^{n} \left(-\frac{1}{2} B_j' P' + \frac{1}{2} P' B_j' \right) \dot{q}_j. \tag{4.35}$$

Now, let us define

$$J_i = \frac{1}{2} \left(P' B_i' - B_i' P' \right). \tag{4.36}$$

Then, Eq. (4.35) can be rewritten as

$$\dot{P}' = J \dot{\boldsymbol{q}}, \tag{4.37}$$

where $J = (J_1 \ \cdots \ J_n)$ is the rotor version of the geometric Jacobian matrix given in (4.6). Therefore, identities (4.35) and (4.37) define the differential kinematics of arbitrary serial robots.

4.4 Inverse Kinematics

As introduced in Sect. 4.1, inverse kinematics is one of the most important problems in robot kinematics.

In this section, Pieper's theorem [81] is revisited using the rich language given by geometric algebra. As pointed out above, this theorem provides a constructive proof of the resolvability of the inverse kinematics of certain classes of serial robots. In practice, however, it is easier to develop a particular geometric strategy for solving the inverse kinematics rather than applying directly Pieper's method. Because of that, the second part of this section is devoted to the development of one of these strategies through an illustrative example. Readers interested in a further analysis of serial manipulators with spherical wrist are referred to [100], where several geometric strategies based on conformal geometric algebra are defined to develop a complete solution for the inverse kinematics of manipulators of this kind.

The starting point is identity (4.14) with $n = 6$. If the desired end-effector pose is denoted by

$$T = \begin{pmatrix} R & \boldsymbol{p} \\ 0 & 1 \end{pmatrix},$$

then it is possible to recover the rotor M that describes such pose by transforming the matrix representation of the world frame T_0 into T, where

$$T_0 = \begin{pmatrix} I_3 & 0 \\ 0 & 1 \end{pmatrix}.$$

First, the position vector $\boldsymbol{p}$ defines the translation

$$\boldsymbol{T}_p = 1 + \|\boldsymbol{p}\| \frac{\boldsymbol{e}_\infty \boldsymbol{p}}{2}. \tag{4.38}$$

Now, to obtain the rotor that relates the orientations defined by the rotation matrices R and I_3, the following procedure is introduced. Both orientations are seen as two sets of vectors in the three-dimensional space: $\{\boldsymbol{e}_1, \boldsymbol{e}_2, \boldsymbol{e}_3\}$ and $\{\boldsymbol{f}_1, \boldsymbol{f}_2, \boldsymbol{f}_3\}$. Here, the vectors of each basis are orthonormal, but, in general, this is not a requirement. Since there is a rotor relating both sets of vectors, the following identity holds:

$$\boldsymbol{f}_k = R\boldsymbol{e}_k\widetilde{R} \quad \text{for } k = 1, 2, 3, \tag{4.39}$$

where a simple expression for rotor R is sought. To obtain such expression, some preliminary results are needed.

As defined in Sect. 3.3, every frame $\{\boldsymbol{e}_1, \boldsymbol{e}_2, \boldsymbol{e}_3\}$ has associated a reciprocal frame $\{\boldsymbol{e}^1, \boldsymbol{e}^2, \boldsymbol{e}^3\}$. Two important properties of reciprocal frames are the following.

4.4.1 (First Property)

$$\boldsymbol{e}_1\boldsymbol{e}^1 + \boldsymbol{e}_2\boldsymbol{e}^2 + \boldsymbol{e}_3\boldsymbol{e}^3 = 3. \tag{4.40}$$

4.4.2 (Second Property)
Given an m-vector A_m,

$$\sum_{k=1}^{3} \boldsymbol{e}_k A_m \boldsymbol{e}^k = (-1)^m (3 - 2m) A_m. \tag{4.41}$$

Readers interested in the proof of these two properties are referred to [29]. Now, rotor R can be written as

$$R = \exp\left(-\frac{\theta B}{2}\right) = \alpha - \beta B, \tag{4.42}$$

where α and β are defined as in the Euler's spinorial formula (see 1.3.9). Its reverse is

$$\widetilde{R} = \exp\left(\frac{\theta B}{2}\right) = \alpha + \beta B. \tag{4.43}$$

Therefore

$$\begin{aligned}\sum_{k=1}^{3} e_k \widetilde{R} e^k &= \sum_{k=1}^{3} e_k(\alpha + \beta B)e^k = \sum_{k=1}^{3} e_k e^k \alpha + \sum_{k=1}^{3} e_k \beta B e^k \\ &\overset{(1)}{=} 3\alpha + (-1)^2(3-4)\beta B = 3\alpha - \beta B \\ &= 3\alpha + \alpha - \alpha - \beta B = 4\alpha - \widetilde{R},\end{aligned} \tag{4.44}$$

where (1) is the result of the application of Eqs. (4.40) and (4.41). Now, merging (4.39) and (4.44), the following expression is obtained:

$$\sum_{k=1}^{3} f_k e^k = \sum_{k=1}^{3} R e_k \widetilde{R} e^k = 4\alpha R - 1. \tag{4.45}$$

It follows that R is a scalar multiple of $1 + \sum_{k=1}^{3} f_k e^k$ and the following formula is established:

$$R = \frac{1 + f_1 e^1 + f_2 e^2 + f_3 e^3}{|1 + f_1 e^1 + f_2 e^2 + f_3 e^3|}. \tag{4.46}$$

Finally, the rotor M describing the transformation between T_0 and T is the product of rotors (4.38) and (4.46), given by

$$M = \boldsymbol{T_p} R, \tag{4.47}$$

and it is, by relation (4.14), equal to

$$M = M_1 M_2 M_3 M_4 M_5 M_6. \tag{4.48}$$

Now, we are in conditions to prove Pieper's theorem using conformal geometric algebra.

4.4.3 (Pieper) *The inverse kinematics of any serial manipulator of 6 DoF with three consecutive joints whose axes are either parallel or intersect at a single point can be analytically solved.*

Proof Let us suppose that three consecutive joint axes intersect at a single point or are parallel. Without loss of generality, it can be assumed that the last three joints (4, 5, and 6) have this property. The remaining cases are left as an exercise for the reader since they are completely analogous to the case developed in this chapter.

Equation (4.48) can be rearranged to isolate the three joints whose axes intersect at a single point:

$$M_4 M_5 M_6 = \widetilde{M}_3 \widetilde{M}_2 \widetilde{M}_1 M. \tag{4.49}$$

Since the axes of joints 4, 5, and 6 intersect at a common point $\boldsymbol{p}$ (the wrist center point), it verifies that

$$M_4 M_5 M_6 \boldsymbol{p} \widetilde{M}_6 \widetilde{M}_5 \widetilde{M}_4 = \boldsymbol{p} = \widetilde{M}_3 \widetilde{M}_2 \widetilde{M}_1 M \boldsymbol{p} \widetilde{M} M_1 M_2 M_3. \tag{4.50}$$

This equation shows how the problem is split into two subproblems

$$\widetilde{M}_1 M \boldsymbol{p} \widetilde{M} M_1 = M_2 M_3 \boldsymbol{p} \widetilde{M}_3 \widetilde{M}_2 \tag{4.51}$$

and

$$M_4 M_5 M_6 = \widetilde{M}_3 \widetilde{M}_2 \widetilde{M}_1 M. \tag{4.52}$$

The first of these equations only involves the joint variables θ_1, θ_2, and θ_3. Once this equation has been solved, by evaluating and solving the second equation, the remaining joint variables, θ_4, θ_5, and θ_6, can be obtained.

Equation (4.51) can be seen as a relation between points:

$$\begin{aligned} \boldsymbol{p}_a &= M_3 \boldsymbol{p} \widetilde{M}_3, \\ \boldsymbol{p}_b &= M_2 \boldsymbol{p}_a \widetilde{M}_2. \end{aligned} \tag{4.53}$$

If the point $\boldsymbol{p}_a$ is rotated around the second joint axis, the plane containing $\boldsymbol{p}_a$ and normal to such axis is invariant. Therefore, it is a scalar multiple of the expression $\boldsymbol{p}_a \widetilde{\ell}_2 + \ell_2 \widetilde{\boldsymbol{p}}_a$, where:

- $\widetilde{\boldsymbol{p}}_a$ does not denote the reverse of $\boldsymbol{p}_a$ but the point obtained as $\widetilde{M}_3 \boldsymbol{p} M_3$ (in fact, the reverse of a vector $\boldsymbol{a} \in \mathcal{G}_3$ verifies $\widetilde{\boldsymbol{a}} = \boldsymbol{a}$);
- ℓ_2 denotes the line defined by the joint axis z_2.

An explanation of these and related properties can be found in [86]. Therefore:

$$\begin{aligned} \boldsymbol{p}_a \widetilde{\ell}_2 + \ell_2 \widetilde{\boldsymbol{p}}_a &= M_3 \boldsymbol{p} \widetilde{M}_3 \widetilde{\ell}_2 + \ell_2 M_3 \boldsymbol{p} \widetilde{M}_3 \\ &\overset{(1)}{=} M_3 \boldsymbol{p} \widetilde{M}_3 \widetilde{M}_2 \widetilde{\ell}_2 M_2 + \widetilde{M}_2 \ell_2 M_2 M_3 \boldsymbol{p} \widetilde{M}_3 \\ &= M_2 M_3 \boldsymbol{p} \widetilde{M}_3 \widetilde{M}_2 \widetilde{\ell}_2 + \ell_2 M_2 M_3 \boldsymbol{p} \widetilde{M}_3 \widetilde{M}_2 \\ &\overset{(2)}{=} \widetilde{M}_1 M \boldsymbol{p} \widetilde{M} M_1 \widetilde{\ell}_2 + \ell_2 \widetilde{M}_1 M \boldsymbol{p} \widetilde{M} M_1, \end{aligned} \tag{4.54}$$

where (1) uses the identity $\ell_2 = \widetilde{M}_2 \ell_2 M_2$ and (2) uses Eq. (4.51). Now, in Eq. (4.54), the second joint variable has been eliminated and, thus, it only depends on θ_1 and θ_3. The process is repeated in order to eliminate another joint variable. From (4.54), the following identity is extracted:

$$M_3 \boldsymbol{p} \widetilde{M}_3 = \widetilde{M}_1 M \boldsymbol{p} \widetilde{M} M_1. \tag{4.55}$$

Again, by rotating $\boldsymbol{p}$ around the third joint axis, the plane containing $\boldsymbol{p}$ and normal to ℓ_3 is defined. Such plane is invariant and is given as a scalar multiple of $\boldsymbol{p}\widetilde{\ell}_3 + \ell_3 \boldsymbol{p}$. Thus

$$\begin{aligned}\boldsymbol{p}\widetilde{\ell}_3 + \ell_3 \boldsymbol{p} &= \boldsymbol{p}\widetilde{M}_3 \widetilde{\ell}_3 M_3 + \widetilde{M}_3 \ell_3 M_3 \boldsymbol{p} \\ &= M_3 \boldsymbol{p} \widetilde{M}_3 \widetilde{\ell}_3 + \ell_3 M_3 \boldsymbol{p} \widetilde{M}_3 \\ &= \widetilde{M}_1 M \boldsymbol{p} \widetilde{M} M_1 \widetilde{\ell}_3 + \ell_3 \widetilde{M}_1 M \boldsymbol{p} \widetilde{M} M_1\end{aligned} \tag{4.56}$$

and, hence, Eq. (4.56) only depends on θ_1. Therefore

$$\left.\begin{aligned}\boldsymbol{p}\widetilde{\ell}_3 + \ell_3 \boldsymbol{p} &= \widetilde{M}_1 M \boldsymbol{p} \widetilde{M} M_1 \widetilde{\ell}_3 + \ell_3 \widetilde{M}_1 M \boldsymbol{p} \widetilde{M} M_1 \\ 1 &= \cos^2(\theta_1) + \sin^2(\theta_1)\end{aligned}\right\} \tag{4.57}$$

is a system of non-linear equations, that in general, has two distinct solutions. Once θ_1 is known, the system

$$\left.\begin{aligned}\boldsymbol{p}_a \widetilde{\ell}_2 + \ell_2 \widetilde{\boldsymbol{p}}_a &= \widetilde{M}_1 M \boldsymbol{p} \widetilde{M} M_1 \widetilde{\ell}_2 + \ell_2 \widetilde{M}_1 M \boldsymbol{p} \widetilde{M} M_1 \\ 1 &= \cos^2(\theta_3) + \sin^2(\theta_3)\end{aligned}\right\} \tag{4.58}$$

can be solved for θ_3. Again, two distinct solutions are obtained. Finally, having found θ_1 and θ_3, θ_2 is calculated using the original Eq. (4.51). For this case, a unique solution is derived.

Once θ_1, θ_2, and θ_3 have been found, they are evaluated in Eq. (4.52), obtaining

$$M_4 M_5 M_6 = N, \tag{4.59}$$

where $N = \widetilde{M}_3 \widetilde{M}_2 \widetilde{M}_1 M$ is a constant rotor. Notice, however, that there are four possible values that N can take corresponding to the four sets of solutions for θ_1, θ_2, and θ_3. For each one of these solutions, rotor M_6 can be eliminated easily acting N over ℓ_6 as follows:

$$N \ell_6 \widetilde{N} = M_4 M_5 M_6 \ell_6 \widetilde{M}_6 \widetilde{M}_5 \widetilde{M}_4 = M_4 M_5 \ell_6 \widetilde{M}_5 \widetilde{M}_4. \tag{4.60}$$

Now, a relation between lines is obtained, since $\ell_a = M_5\ell_6\widetilde{M}_5$ and $\ell_b = M_4\ell_a\widetilde{M}_4$ for two lines ℓ_a and ℓ_b. If ℓ_a is rotated around the fourth joint axis, an invariant plane is generated. Such plane is a multiple scalar of $\ell_a\widetilde{\ell}_4 + \ell_4\widetilde{\ell}_a$, which leads to:

$$\begin{aligned} \ell_a\widetilde{\ell}_4 + \ell_4\widetilde{\ell}_a &= M_5\ell_6\widetilde{M}_5\widetilde{\ell}_4 + \ell_4 M_5\widetilde{\ell}_6\widetilde{M}_5 \\ &= M_5\ell_6\widetilde{M}_5\widetilde{M}_4\widetilde{\ell}_4 M_4 + \widetilde{M}_4\ell_4 M_4 M_5\widetilde{\ell}_6\widetilde{M}_5 \\ &= M_4 M_5\ell_6\widetilde{M}_5\widetilde{M}_4\widetilde{\ell}_4 + \ell_4 M_4 M_5\widetilde{\ell}_6\widetilde{M}_5\widetilde{M}_4 \\ &= N\ell_6\widetilde{N}\widetilde{\ell}_4 + \ell_4 N\widetilde{\ell}_6\widetilde{N}\ell_4. \end{aligned} \tag{4.61}$$

Since Eq. (4.61) only depends on θ_5, by adding the equation $\cos^2(\theta_5)+\sin^2(\theta_5) = 1$ to (4.61), two distinct solutions are derived.

Once θ_5 has been calculated, the following identity is used for solving θ_4:

$$M_4 M_5\ell_6\widetilde{M}_5\widetilde{M}_4 = N\ell_6\widetilde{N}. \tag{4.62}$$

Equation (4.62) has a unique solution. Finally, the original Eq. (4.59) is used to recover θ_6. □

For a given end-effector pose, it has been shown that the robots of this kind have a maximum of eight distinct solutions for the inverse kinematics. In the general case, where no three consecutive joint axes intersect or are parallel, it can be proven that the inverse kinematics of 6 DoF serial robots has up to 16 distinct solutions.

As commented above, this proof provides the different equations required for solving the inverse kinematics. However, a geometric strategy using the conformal model of the spatial geometric algebra is usually more efficient and easier to formulate.

The second part of this section is focused on the development of one of these strategies. For that purpose, an illustrative example is introduced. Stäubli TX90 is a 6 DoF manipulator with a spherical wrist (see Fig. 4.4). Thus, by translating the end-effector position to the wrist center point, it is assured that the first three joints contribute to the position and orientation but the last three only contribute to the orientation. Since this transformation is fixed, i.e., it does not depend on any joint variable, the inverse kinematics can be solved with the wrist center point $\boldsymbol{p}_w$ as target position.

The first part of the strategy consists of solving the inverse position problem, i.e., finding the values of the joint variables needed for obtaining the target position. Since only the first three joints contribute to the position, it is enough to calculate θ_1, θ_2, and θ_3. Let us denote by p_0 the null vector representation of the point placed at the origin of the world frame. Such representation is the result of applying the Hestenes' map to $\boldsymbol{e}_0$, according to 2.1.4. Therefore, the intermediate points p_1 and p_2 are needed to be found.

Fig. 4.4 Stäubli TX90: initial position

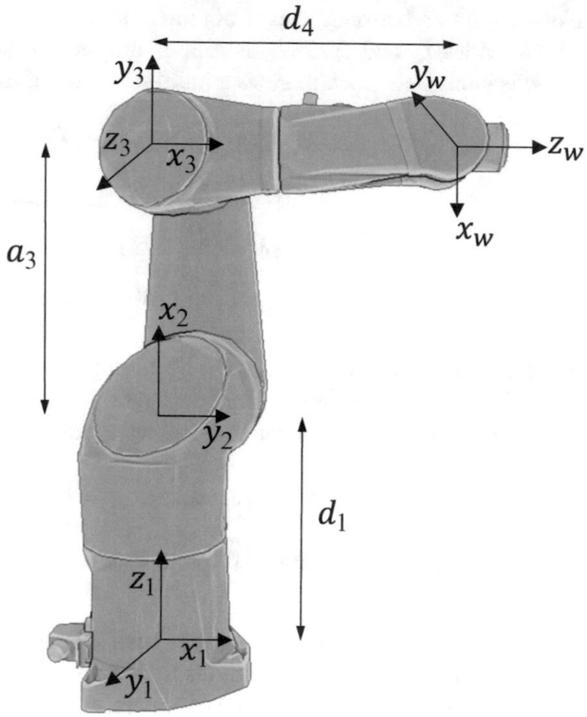

Point p_0 is translated in the direction of z_1 and, thus, $p_1 = d_1\boldsymbol{e}_1 + 1/2d_1^2\boldsymbol{e}_\infty + \boldsymbol{e}_0$ where, again, its representation comes from 2.1.4.

Once the point p_1 has been obtained, the point p_2 is established by intersecting two spheres and one plane as shown in Fig. 4.5a, b. Planes are constructed in conformal geometric algebra with three non-collinear points (see Theorem 2.3.5), while spheres need four points (see Theorem 2.3.6):

$$\pi_1 = p_0 \wedge p_1 \wedge p_w \wedge \boldsymbol{e}_\infty, \tag{4.63}$$

where p_w is the null vector representation of point $\boldsymbol{p}_w$. On the other side, the two spheres are defined as:

$$S_1^* = \left(p_1 - \frac{1}{2}a_3^2\boldsymbol{e}_\infty\right), \tag{4.64}$$

$$S_2^* = \left(p_w - \frac{1}{2}d_4^2\boldsymbol{e}_\infty\right), \tag{4.65}$$

where the inner representation of both spheres has been chosen because it only needs their center and radius.

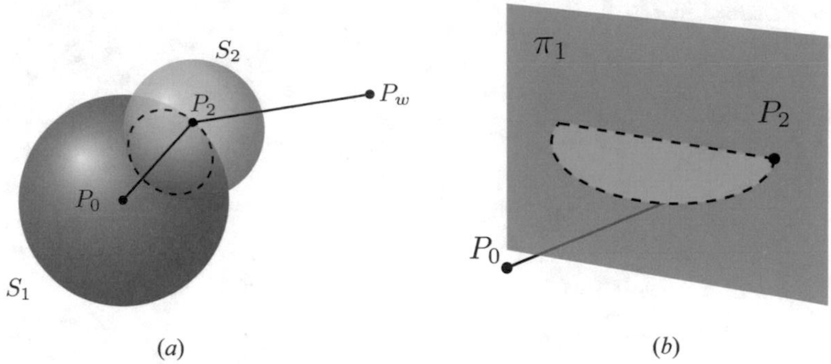

Fig. 4.5 Computation of P_2: (**a**) P_2 in the intersection of two spheres. (**b**) P_2 in the intersection of a circle and a plane

Now, the intersection between these three geometric entities is computed. For that, a definition of such intersection in conformal geometric algebra is required.

4.4.4 *The intersection or meet between two geometric entities O_1 and O_2 is defined as the element of $\mathcal{C}$:*

$$O_1 \vee O_2 = \left[O_1^* \wedge O_2^*\right]^* .$$

As shown in Exercise E.4.6, the intersection of the plane (4.63) with the spheres (4.64) and (4.65) is a bivector:

$$B_2 = S_1 \vee S_2 \vee \pi_1. \tag{4.66}$$

This bivector represents a pair of points in the conformal geometric algebra, so $B_2 = b_1 \wedge b_2$ for some null points b_1 and b_2. To extract such points from (4.66), the following equations are used:

$$\begin{aligned} b_1 &= -\widetilde{P}[(b_1 \wedge b_2) \cdot \boldsymbol{e}_\infty]P, \\ b_2 &= P[(b_1 \wedge b_2) \cdot \boldsymbol{e}_\infty]\widetilde{P}, \end{aligned} \tag{4.67}$$

where P denotes the projector operator defined as:

$$P = \frac{1}{2}\left(1 + \frac{b_1 \wedge b_2}{|b_1 \wedge b_2|}\right). \tag{4.68}$$

Clearly, p_2 is equal to one of the recovered points b_i (for $i = 1, 2$).

It only remains to find the joint variables. Since the first three joints are revolute, their joint variables are angles. First, we need to construct three auxiliary lines with the already obtained points:

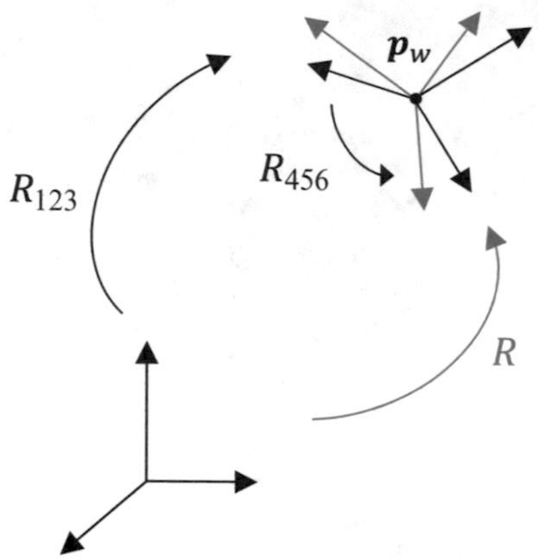

Fig. 4.6 Relation between the orientations

$$\ell_1 = p_0 \wedge p_1 \wedge \boldsymbol{e}_\infty, \tag{4.69}$$

$$\ell_2 = p_1 \wedge p_2 \wedge \boldsymbol{e}_\infty, \tag{4.70}$$

$$\ell_3 = p_2 \wedge p_w \wedge \boldsymbol{e}_\infty. \tag{4.71}$$

Finally, using the geometric entities (4.63), (4.69)–(4.71) and Exercise E.2.2, the joint variables are calculated:

$$\theta_1 = \angle(e_1, \pi_1), \tag{4.72}$$

$$\theta_2 = \angle(\ell_1, \ell_2), \tag{4.73}$$

$$\theta_3 = \angle(\ell_2, \ell_3), \tag{4.74}$$

where $\angle(\cdot, \cdot)$ denotes the main angle defined by the two geometric entities.

The second part consists of solving the inverse orientation problem. Since $\theta_1, \theta_2, \theta_3$ are known, it only remains to find θ_4, θ_5, and θ_6. With $\theta_1, \theta_2, \theta_3$, the rotor defining the orientation of the wrist center point $\boldsymbol{p}_w$ under the effect of these joints can be calculated. Let us denote by R_{123} such rotor. Besides, recall that R denotes the rotor that relates the orientation of the world frame with the orientation of the end-effector. Then, the rotor that defines the rotation between R_{123} and R can be obtained from the formula

$$R = R_{123} R_{456}, \tag{4.75}$$

where the rotor R_{456} only depends on θ_4, θ_5, and θ_6 (see Fig. 4.6).

Now, the idea is to split R_{456} into three different rotors as follows:

$$R_{456} = R_4 R_5 R_6, \tag{4.76}$$

where $R_i = \cos(\theta_i/2) - \sin(\theta_i/2) B_i$ with angle of rotation θ_i and bivector B_i. The angles θ_4, θ_5, and θ_6 will correspond to the desired joint variables.

If Eq. (4.76) is expanded, the following expression is obtained:

$$\left(\cos\left(\frac{\theta_4}{2}\right)-\sin\left(\frac{\theta_4}{2}\right)B_4\right)\left(\cos\left(\frac{\theta_5}{2}\right)-\sin\left(\frac{\theta_5}{2}\right)B_5\right)\left(\cos\left(\frac{\theta_6}{2}\right)-\sin\left(\frac{\theta_6}{2}\right)B_6\right)$$
$$= \cos\left(\frac{\theta_4}{2}\right)\cos\left(\frac{\theta_5}{2}\right)\cos\left(\frac{\theta_6}{2}\right) - \cos\left(\frac{\theta_4}{2}\right)\cos\left(\frac{\theta_6}{2}\right)\sin\left(\frac{\theta_5}{2}\right)B_5$$
$$- \cos\left(\frac{\theta_5}{2}\right)\cos\left(\frac{\theta_6}{2}\right)\sin\left(\frac{\theta_4}{2}\right)B_4 - \cos\left(\frac{\theta_4}{2}\right)\cos\left(\frac{\theta_5}{2}\right)\sin\left(\frac{\theta_6}{2}\right)B_6$$
$$+ \cos\left(\frac{\theta_6}{2}\right)\sin\left(\frac{\theta_4}{2}\right)\sin\left(\frac{\theta_5}{2}\right)B_4B_5 + \cos\left(\frac{\theta_4}{2}\right)\sin\left(\frac{\theta_5}{2}\right)\sin\left(\frac{\theta_6}{2}\right)B_5B_6$$
$$+ \cos\left(\frac{\theta_5}{2}\right)\sin\left(\frac{\theta_4}{2}\right)\sin\left(\frac{\theta_6}{2}\right)B_4B_6 - \sin\left(\frac{\theta_4}{2}\right)\sin\left(\frac{\theta_5}{2}\right)\sin\left(\frac{\theta_6}{2}\right)B_4B_5B_6, \tag{4.77}$$

where, besides, we can express R_{456} as follows:

$$R_{456} = \cos\left(\frac{\theta}{2}\right) - \sin\left(\frac{\theta}{2}\right)B_{456}, \tag{4.78}$$

for an angle θ and a bivector B_{456}. As it can be observed, working directly with Eq. (4.77) is highly difficult, so an alternative path is required.

4.4.5 *Let us denote by R a rotor in the conformal geometric algebra $\mathcal{C}$. Then, $RH(\boldsymbol{x})\widetilde{R} = H(R\boldsymbol{x}\widetilde{R})$ for every $\boldsymbol{x} \in \mathbb{R}^3$, where $H(\cdot)$ denotes the Hestenes' map* (*see* 2.1.4).

Proof It follows immediately from applying 2.4.2 to rotors. □

Therefore, we can define the rotations in $\mathcal{C}$ using the bivectors of $\mathcal{G}_3$ and, as a result, B_{456} can be expressed as a linear combination with respect to the basis bivectors of $\mathcal{G}_3$:

$$B_{456} = \beta_1\boldsymbol{e}_{23} + \beta_2\boldsymbol{e}_{13} + \beta_3\boldsymbol{e}_{12}, \tag{4.79}$$

for some $\beta_i \in \mathbb{R}$. Now, Eq. (4.78) can be rewritten as

$$R_{456} = \cos\left(\frac{\theta}{2}\right) - \sin\left(\frac{\theta}{2}\right)\beta_1\boldsymbol{e}_{23} - \sin\left(\frac{\theta}{2}\right)\beta_2\boldsymbol{e}_{13} - \sin\left(\frac{\theta}{2}\right)\beta_3\boldsymbol{e}_{12}. \tag{4.80}$$

As it happens with the *Euler angles*, different conventions can be adopted. Depending on the chosen convention, a particular set of equations is obtained. For example, if

$$B_4 = \boldsymbol{e}_{23}, \quad B_5 = \boldsymbol{e}_{13}, \quad B_6 = \boldsymbol{e}_{12} \tag{4.81}$$

that correspond to the Euler angles convention XYZ, then:

$$B_4B_5 = B_6;\ B_5B_6 = B_4;\ B_4B_6 = -B_5;\ B_4B_5B_6 = -1. \tag{4.82}$$

Hence, by regrouping the terms of (4.77) and equating them to the terms of (4.80), the following system of equations is obtained:

$$\begin{aligned}
\beta_1' &= \cos\left(\frac{\theta_4}{2}\right)\sin\left(\frac{\theta_5}{2}\right)\sin\left(\frac{\theta_6}{2}\right) - \cos\left(\frac{\theta_5}{2}\right)\cos\left(\frac{\theta_6}{2}\right)\sin\left(\frac{\theta_4}{2}\right),\\
\beta_2' &= -\cos\left(\frac{\theta_4}{2}\right)\cos\left(\frac{\theta_6}{2}\right)\sin\left(\frac{\theta_5}{2}\right) - \cos\left(\frac{\theta_5}{2}\right)\sin\left(\frac{\theta_4}{2}\right)\sin\left(\frac{\theta_6}{2}\right),\\
\beta_3' &= \cos\left(\frac{\theta_6}{2}\right)\sin\left(\frac{\theta_4}{2}\right)\sin\left(\frac{\theta_5}{2}\right) - \cos\left(\frac{\theta_4}{2}\right)\cos\left(\frac{\theta_5}{2}\right)\sin\left(\frac{\theta_6}{2}\right),
\end{aligned} \tag{4.83}$$

where $\beta_i' = -\sin(\theta/2)\beta_i$. This is a system of non-linear equations in θ_4, θ_5, and θ_6 that can be solved analytically. However, (4.83) is difficult to solve. By changing the convention, a more suitable system of equations can be obtained. Therefore, by setting

$$B_4 = \boldsymbol{e}_{12}, \quad B_5 = \boldsymbol{e}_{13}, \text{ and } B_6 = \boldsymbol{e}_{12}, \tag{4.84}$$

the following relations hold:

$$B_4B_5 = -\boldsymbol{e}_{23};\ B_5B_6 = \boldsymbol{e}_{23};\ B_4B_6 = -1;\ B_4B_5B_6 = B_5. \tag{4.85}$$

Then, the system of equations obtained is:

$$\begin{aligned}
\beta_1' &= \cos\left(\frac{\theta_4}{2}\right)\sin\left(\frac{\theta_5}{2}\right)\sin\left(\frac{\theta_6}{2}\right) - \cos\left(\frac{\theta_6}{2}\right)\sin\left(\frac{\theta_4}{2}\right)\sin\left(\frac{\theta_5}{2}\right),\\
\beta_2' &= -\cos\left(\frac{\theta_4}{2}\right)\cos\left(\frac{\theta_6}{2}\right)\sin\left(\frac{\theta_5}{2}\right) - \sin\left(\frac{\theta_4}{2}\right)\sin\left(\frac{\theta_5}{2}\right)\sin\left(\frac{\theta_6}{2}\right),\\
\beta_3' &= -\cos\left(\frac{\theta_5}{2}\right)\cos\left(\frac{\theta_6}{2}\right)\sin\left(\frac{\theta_4}{2}\right) - \cos\left(\frac{\theta_4}{2}\right)\cos\left(\frac{\theta_5}{2}\right)\sin\left(\frac{\theta_6}{2}\right),
\end{aligned} \tag{4.86}$$

that can be simplified as follows:

$$\beta_1' = \sin\left(\frac{\theta_6 - \theta_4}{2}\right)\sin\left(\frac{\theta_5}{2}\right), \tag{4.87}$$

$$\beta_2' = -\cos\left(\frac{\theta_6 - \theta_4}{2}\right)\sin\left(\frac{\theta_5}{2}\right), \tag{4.88}$$

$$\beta_3' = -\sin\left(\frac{\theta_6 + \theta_4}{2}\right)\cos\left(\frac{\theta_5}{2}\right). \tag{4.89}$$

So, by squaring and adding (4.87) and (4.88), we obtain

$$\begin{aligned}(\beta')_1^2 + (\beta')_2^2 &= \sin^2\left(\frac{\theta_6 - \theta_4}{2}\right)\sin^2\left(\frac{\theta_5}{2}\right) + \cos^2\left(\frac{\theta_6 - \theta_4}{2}\right)\sin^2\left(\frac{\theta_5}{2}\right) = \\ &= \sin^2\left(\frac{\theta_5}{2}\right),\end{aligned} \tag{4.90}$$

where the solution of this trigonometric equation is

$$\theta_5 = \frac{\sin^{-1}\left(\pm\sqrt{(\beta')_1^2 + (\beta')_2^2}\right)}{2}. \tag{4.91}$$

Finally, if $\sin(\theta_5/2) \neq 0$

$$\beta_1'' = \frac{\beta_1'}{\sin(\theta_5/2)} \quad \beta_2'' = \frac{-\beta_2'}{\sin(\theta_5/2)} \quad \beta_3'' = \frac{-\beta_3'}{\cos(\theta_5/2)}, \tag{4.92}$$

and, therefore:

$$\begin{aligned}\beta_1'' &= \sin\left(\frac{\theta_6 - \theta_4}{2}\right) & & \frac{\theta_6 - \theta_4}{2} = \sin^{-1}(\beta_1''), \\ & & \Longrightarrow & \\ \beta_3'' &= \sin\left(\frac{\theta_6 + \theta_4}{2}\right) & & \frac{\theta_6 + \theta_4}{2} = \sin^{-1}(\beta_3''),\end{aligned} \tag{4.93}$$

that is solved easily as follows:

$$\begin{aligned}\theta_6 &= \sin^{-1}(\beta_1'') + \sin^{-1}(\beta_3''), \\ \theta_4 &= \sin^{-1}(\beta_3'') - \sin^{-1}(\beta_1'').\end{aligned} \tag{4.94}$$

For the case where $\sin(\theta_5/2) = 0$, an alternative solution, also based on the use of rotors, can be found in [100].

This completes the resolution of the inverse orientation problem and, therefore, the description of the geometric strategy presented in this section. As it has been seen, this method strongly depends on the geometry of the robot but it is much easier to formulate and solve than Pieper's method. Besides, it allows a better understanding of the geometry of the robot. If implemented, it has less computational cost than any algorithm based on the geometric Jacobian matrix and, as a result, it turns to be more efficient for computing the inverse kinematics. Again, conformal geometric algebra has proven to be a powerful tool for simplifying the formulation and resolution of robotic problems.

4.5 Identification of Singularities

Although identity (4.37) in Sect. 4.3 provides a computationally efficient version of the geometric Jacobian matrix based on rotors, $J = (J_1, \ldots, J_n)$ is still a matrix. Geometric algebra provides a framework where this problem is formulated and solved in an easy and compact way. For that purpose, only 6 DoF serial robots are considered.

The more natural way of formulating the singularity problem is through the six-dimensional geometric algebra $\mathcal{G}_6$, that extends the three-dimensional algebra $\mathcal{G}_3$ introduced in Sect. 1.3.

4.5.1 *A line ℓ can be fully specified by two three-dimensional vectors: its direction vector $\boldsymbol{v}$ and its position vector $\boldsymbol{p}$. The Plücker coordinates of ℓ define a six-dimensional vector $[\boldsymbol{v}\ \boldsymbol{v} \times \boldsymbol{p}]^T$, where $\boldsymbol{v} \times \boldsymbol{p}$ denotes the moment vector of ℓ.*

The identity that allows the computation of the singularities of a given 6 DoF serial manipulator is the following:

4.5.2 *If $S_i(\boldsymbol{q})$ denotes the six-dimensional vector whose components are the Plücker coordinates of the i-th screw axis, then:*

$$S_1(\boldsymbol{q}) \wedge \cdots \wedge S_6(\boldsymbol{q}) = \det(S_1(\boldsymbol{q}) \ \cdots \ S_6(\boldsymbol{q}))\boldsymbol{e}_1 \wedge \cdots \wedge \boldsymbol{e}_6. \tag{4.95}$$

Proof Let us start with two vectors of $\mathcal{G}_2$. Let $\boldsymbol{a}_1, \boldsymbol{a}_2 \in \mathcal{G}_2$ be such vectors. Therefore, $\boldsymbol{a}_1 = a_{11}\boldsymbol{e}_1 + a_{12}\boldsymbol{e}_2 = (a_{11}, a_{12})$ and $\boldsymbol{a}_2 = a_{21}\boldsymbol{e}_1 + a_{22}\boldsymbol{e}_2 = (a_{21}, a_{22})$. The exterior product of both vectors is computed as follows:

$$\begin{aligned} \boldsymbol{a}_1 \wedge \boldsymbol{a}_2 &= (a_{11}\boldsymbol{e}_1 + a_{12}\boldsymbol{e}_2) \wedge (a_{21}\boldsymbol{e}_1 + a_{22}\boldsymbol{e}_2) \\ &= (a_{11}a_{22} - a_{21}a_{12})\boldsymbol{e}_1 \wedge \boldsymbol{e}_2 \\ &= \det(\boldsymbol{a}_1\ \boldsymbol{a}_2)\boldsymbol{e}_1 \wedge \boldsymbol{e}_2. \end{aligned} \tag{4.96}$$

Clearly, (4.96) can be extended to a set of m vectors of $\mathcal{G}_m$:

$$\boldsymbol{a}_1 \wedge \cdots \wedge \boldsymbol{a}_m = \det(\boldsymbol{a}_1 \ \cdots \ \boldsymbol{a}_m)\boldsymbol{e}_1 \wedge \cdots \wedge \boldsymbol{e}_m. \tag{4.97}$$

Since we have only made use of the exterior product, this result is also true for any exterior algebra $\wedge E_m$. In particular, Eq. (4.97) is true for any set of six vectors $\boldsymbol{a}_1, \cdots, \boldsymbol{a}_6$ of $\mathcal{G}_6$, which gives the desirable identity. □

Taking the dual of 4.5.2, an expression for computing the singularities is obtained:

$$(S_1(\boldsymbol{q}) \wedge \cdots \wedge S_6(\boldsymbol{q}))^* = 0, \tag{4.98}$$

where, clearly, the solution of this equation are the singularities of the serial robot whose axes can be expressed in Plücker coordinates as $S_i(\boldsymbol{q})$.

In general, to obtain the singularities is difficult since Eq. (4.98) is non-linear. However, for robots with a spherical wrist a simplification can be made. As mentioned in Sect. 4.1, the singularities of manipulators of this kind can be decoupled into position and orientation singularities.

For each screw axis in Plücker coordinates $S_i(\boldsymbol{q}) = (\boldsymbol{v}_i(\boldsymbol{q}), \boldsymbol{m}_i(\boldsymbol{q}))$, the axis and moment vectors can be expressed as vectors in $\mathcal{G}_3$. Again, by (4.97) but with $m = 3$, the position singularities are obtained as the solutions of the following equation:

$$(\boldsymbol{m}_1(\boldsymbol{q}) \wedge \boldsymbol{m}_2(\boldsymbol{q}) \wedge \boldsymbol{m}_3(\boldsymbol{q}))^* = 0, \tag{4.99}$$

while the orientation singularities are the solutions of

$$(\boldsymbol{v}_4(\boldsymbol{q}) \wedge \boldsymbol{v}_5(\boldsymbol{q}) \wedge \boldsymbol{v}_6(\boldsymbol{q}))^* = 0. \tag{4.100}$$

As an illustrative example, let us consider the Stäubli TX90 manipulator introduced in Sect. 4.4. Since it has a spherical wrist, the above-mentioned simplification can be made and, as a result, its position singularities are the solutions of

$$\begin{aligned} &\sin(q_3)(9031250 + 76765625\sin(q_2) - 76765625\cos(q_2)\sin(q_3) \\ &\quad + 76765625\cos(q_3)\sin(q_2)) = 0, \end{aligned} \tag{4.101}$$

while its orientation singularities are the solutions of

$$-\sin(q_5) = 0. \tag{4.102}$$

There are two position singularities, namely

- $q_3 = 0$ or π, and
- the values of q_2, q_3 such that $\sin(q_2) + \sin(q_2 - q_3) = -0.118$ if $q_3 \neq 0, \pi$;

and just one orientation singularity, namely $q_5 = 0$ or π.

4.6 Exercises

E.4.1 Consider the planar manipulator introduced in Sect. 4.2 and let $a_1 = a_2 = a_3 = a$. If the initial pose is $(3a, 0, 0)$, obtain the pose associated with the configuration $(\pi/2, \pi/4, -\pi/6)$.

E.4.2 Compute the forward kinematics multivector of the Stäubli TX90 using its D-H parameters (see Fig. 4.7a).

E.4.3 Deduce that the expression for the derivative of a rotor $R(t)$ with respect to time is $\dot{R}(t) = -\frac{1}{2}BR(t)$ where B is the bivector associated with $R(t)$.

	α_i	a_i	d_i	θ_i
1	0	0	480	θ_1
2	$-\frac{\pi}{2}$	0	0	$\theta_2-\frac{\pi}{2}$
3	0	425	0	$\theta_3+\frac{\pi}{2}$
4	$\frac{\pi}{2}$	0	425	θ_4
5	$-\frac{\pi}{2}$	0	0	θ_5
6	$\frac{\pi}{2}$	0	0	θ_6

(a)

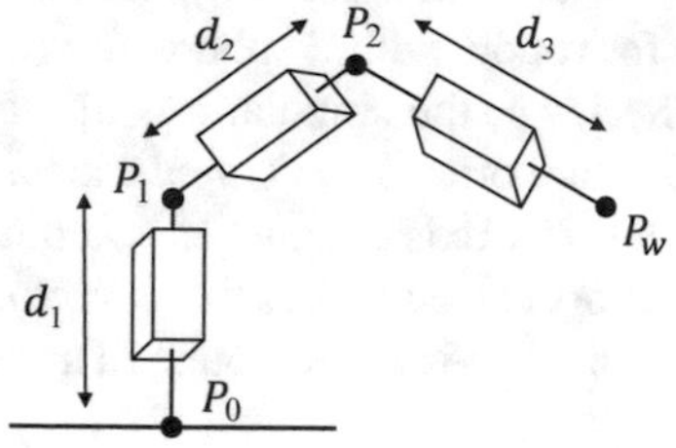

(b)

Fig. 4.7 (**a**) Stäubli TX90 D-H parameters. (**b**) Cartesian manipulator

E.4.4 Derive an equivalent expression to the rate relation (4.37).
Hint: Express the forward kinematics equations in terms of the exponentials of bivectors. Manipulate such equations using the expression of the rotor's derivative.

E.4.5 Prove Pieper's theorem for the case of three consecutive revolute joints with parallel axes.
Hint: The proof follows the one developed in this chapter but considering an invariant plane instead of an invariant point. Such plane is normal to the parallel joint axes.

E.4.6 Show that the meet of two spheres S_1 and S_2 with a plane π is always a bivector.

E.4.7 Consider the Cartesian manipulator depicted in Fig. 4.7b. Its non-null D-H parameters are: $\alpha_2 = \theta_2 = \theta_3 = \dfrac{\pi}{2}$ and $\alpha_3 = -\frac{\pi}{2}$. Solve the inverse position problem for a target position $\boldsymbol{p} = (p_x, p_y, p_z)$.

E.4.8 Formulate the singularity problem for general manipulators in $\mathcal{G}_{3,1}$.
Hint: There are six different basis bivectors in $\mathcal{G}_{3,1}$. Besides, the screw axis of each joint can be seen (in Plücker coordinates) as a bivector.

Chapter 5
Molecular Geometry

The 3D structure of a molecule is fundamental for understanding its function, especially in the case of proteins [28]. The calculation of protein structures can be tackled experimentally, through Nuclear Magnetic Resonance (NMR) spectroscopy and X-ray crystallography [9], or theoretically, via molecular potential energy minimization [56, 57].

This chapter will explain how geometric algebra and conformal geometric algebra can be used to model problems in molecular geometry. In particular, we will focus on problems related to protein structure determination using NMR data.

5.1 Distance Geometry

The X-ray crystallography was the first method applied to the determination of protein structures, where crystallized proteins were considered as rigid objects. The development of NMR experiments allowed to study protein molecules in solution, putting in evidence the internal dynamics of the protein structures [94].

NMR spectroscopy is based on the measurement of distances between hydrogen atoms that are close enough in a protein molecule, and the problem is to determine its 3D structure using this distance information.

We can use two types of sets (the entities V and their relationships E) and a real function d on E to model this problem: V represents the set of atoms, E represents the set of atom pairs for which a distance is available, and the function $d : E \to (0, \infty)$ assigns non-negative real numbers to each pair $\{u, v\} \in E$.

When we consider V, E, d together we have a *weighted graph*, denoted by $G = (V, E, d)$. We say that it is *simple* and *undirected* because, respectively, if $\{u, v\} \in E$ then $u \neq v$, and $\{u, v\} = \{v, u\}$.

C. Lavor et al., *A Geometric Algebra Invitation to Space-Time Physics, Robotics and Molecular Geometry*, SpringerBriefs in Mathematics,
https://doi.org/10.1007/978-3-319-90665-2_5

A graph $G = (V, E, d)$ is just a mathematical abstraction to represent the problem data. The problem itself is to find a function $x : V \to \mathbb{R}^3$ that associates each element of V with a point in $\mathbb{R}^3$ in such a way that the Euclidean distances between the points correspond to the values given by d. This is a *Distance Geometry Problem* (DGP) in $\mathbb{R}^3$, formally described as follows.

5.1.1 (Statement of the DGP) Given a simple undirected graph $G = (V, E, d)$ whose edges are weighted by $d : E \to (0, \infty)$, find a function $x : V \to \mathbb{R}^3$ such that

$$\forall \{u, v\} \in E, \ ||\boldsymbol{x}_u - \boldsymbol{x}_v|| = d_{u,v}, \tag{5.1}$$

where $\boldsymbol{x}_u = x(u)$, $\boldsymbol{x}_v = x(v)$, $d_{u,v} = d(\{u, v\})$, and $||\boldsymbol{x}_u - \boldsymbol{x}_v||$ is the Euclidean distance between $\boldsymbol{x}_u$ and $\boldsymbol{x}_v$.

In addition to protein conformation [23], there are many other applications of the DGP. See [8, 68] for recent surveys, [76] for an edited book with different applications, and [65] for DG historical notes.

Because Eq. (5.1) pose difficulties to be solved numerically, a common approach is to formulate the DGP as an optimization problem,

$$\underset{\boldsymbol{x}_1,\ldots,\boldsymbol{x}_n \in \mathbb{R}^3}{\text{minimize}} \sum_{\{u,v\} \in E} \left(||\boldsymbol{x}_u - \boldsymbol{x}_v||^2 - d_{u,v}^2 \right)^2,$$

where $|V| = n$. In [58], some optimization algorithms have been tested but none of them scale well to medium or large instances. A survey on different methods to the DGP is given in [67].

If we do not consider the effect of translations and rotations, the number of solutions of the DGP depends on the structure of the associated graph $G = (V, E, d)$ [69]. For example, if the set of edges E contains all possible pairs from V (and the solution set is not empty), there is only one solution which can be found in linear time. In general, the problem is NP-hard [84].

Using information from the chemistry of proteins and assuming that the input data are correct and precise, the DGP solution set is finite (up to translations and rotations), allowing the application of combinatorial methods [60].

5.2 Discretizable Molecular Distance Geometry Problem

The NMR data and the chemistry of proteins allow us to define a vertex order $v_1, \ldots, v_n \in V$ such that [15, 35, 75] (we denote $\boldsymbol{x}_i$ instead of $\boldsymbol{x}_{v_i}$ and $d_{i,j}$ instead of $d_{v_i v_j}$):

1. For the first three vertices, there exist $\boldsymbol{x}_1, \boldsymbol{x}_2, \boldsymbol{x}_3 \in \mathbb{R}^3$ satisfying Eq. (5.1);
2. Each vertex with rank greater than three is adjacent to at least three contiguous predecessors, *i.e.*

$$\forall i > 3, \{v_{i-3}, v_i\}, \{v_{i-2}, v_i\}, \{v_{i-1}, v_i\} \in E.$$

The set of DGP instances with this order is called the *Discretizable Molecular Distance Geometry Problem* (DMDGP), and the order itself is a DMDGP order [59]. To guarantee a finite number of solutions, the DMDGP definition also requires that the *strict* triangular inequality must be satisfied for each three consecutive vertices, *i.e.* $\forall i > 3$,

$$d_{i-3,i-2} + d_{i-2,i-1} > d_{i-3,i-1}. \tag{5.2}$$

Property 1 above guarantees that the DMDGP solution set will not consider solutions modulo rotations and translations. From property 2, the position for vertex v_4 can be obtained solving the quadratic system

$$\begin{aligned}
||\boldsymbol{x}_4 - \boldsymbol{x}_1||^2 &= d_{1,4}^2,\\
||\boldsymbol{x}_4 - \boldsymbol{x}_2||^2 &= d_{2,4}^2,\\
||\boldsymbol{x}_4 - \boldsymbol{x}_3||^2 &= d_{3,4}^2,
\end{aligned}$$

which can result in up to two possible positions for v_4. Using the same idea, for each position already determined for v_4, we obtain other two for v_5, and so on. The DMDGP definition then implies that the search space is finite, having 2^{n-3} possible solutions [59].

In addition to $\{v_{i-3}, v_i\}, \{v_{i-2}, v_i\}, \{v_{i-1}, v_i\} \in E$, for some $i > 3$, we may also have $\{v_j, v_i\} \in E$, $j < i - 3$, adding another equation to the system related to v_i:

$$\begin{aligned}
||\boldsymbol{x}_i - \boldsymbol{x}_j||^2 &= d_{j,i}^2,\\
||\boldsymbol{x}_i - \boldsymbol{x}_{i-3}||^2 &= d_{i-3,i}^2,\\
||\boldsymbol{x}_i - \boldsymbol{x}_{i-2}||^2 &= d_{i-2,i}^2,\\
||\boldsymbol{x}_i - \boldsymbol{x}_{i-1}||^2 &= d_{i-1,i}^2.
\end{aligned}$$

Subtracting one of these equations from the others, we eliminate the term $||\boldsymbol{x}_i||^2$ and obtain a linear system in the variable $\boldsymbol{x}_i$ (the positions $\boldsymbol{x}_j, \boldsymbol{x}_{i-3}, \boldsymbol{x}_{i-2}, \boldsymbol{x}_{i-1} \in \mathbb{R}^3$ were already calculated). If the points $\boldsymbol{x}_j, \boldsymbol{x}_{i-3}, \boldsymbol{x}_{i-2}, \boldsymbol{x}_{i-1}$ are not in the same plane, we have a unique solution $\boldsymbol{x}_i^*$ for v_i, supposing $||\boldsymbol{x}_i^* - \boldsymbol{x}_j|| = d_{j,i}$. It may happen that both possible positions for v_i are infeasible with respect to additional distances. In this case, it is necessary to consider the other possible position for v_{i-1} and repeat the procedure [59].

The idea described above can be formalized in an algorithm called Branch-and-Prune (BP) [66], which can be exponential in the worst case. However, considering precise input data, its performance is impressive in terms of efficiency and reliability [59].

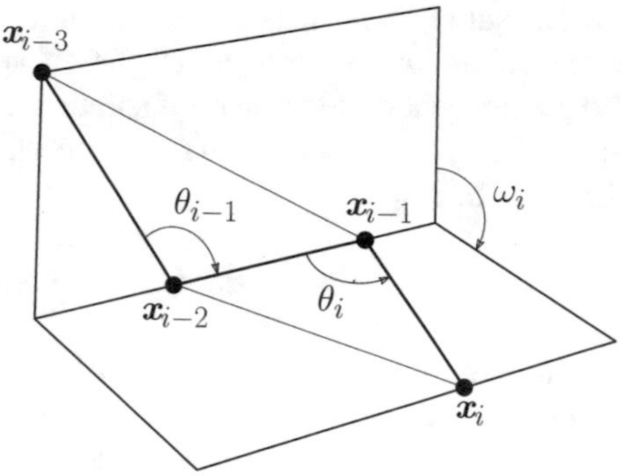

Fig. 5.1 Cartesian and internal coordinates

5.3 Cartesian and Internal Coordinates

A DMDGP order $v_1, \ldots, v_n$ represents bonded atoms of a protein molecule, where a solution of the problem is given by the positions of such vertices in $\mathbb{R}^3$ satisfying the DMDGP equations. In addition to the Cartesian coordinates $\boldsymbol{x}_1, \ldots, \boldsymbol{x}_n \in \mathbb{R}^3$, the molecular 3D structure related to the vertices $v_1, \ldots, v_n$ of a DMDGP instance can also be described in terms of the *internal coordinates* [80], given by the *bond lengths* d_i (the Euclidean distance related to v_{i-1}, v_i), for $i = 2, \ldots, n$, the *bond angles* θ_i (the angle associated to v_{i-2}, v_{i-1}, v_i), for $i = 3, \ldots, n$, and the *torsion angles* ω_i (the angle between the normals through the planes defined by $v_{i-3}, v_{i-2}, v_{i-1}$ and v_{i-2}, v_{i-1}, v_i), for $i = 4, \ldots, n$ (Fig. 5.1).

In protein structure calculations, the internal coordinates are very useful because it is common to assume that all bond lengths and bond angles are fixed at their equilibrium values [28]. This means that all the values d_i, for $i = 2, \ldots, n$, and θ_i, for $i = 3, \ldots, n$, are given *a priori*, and that the 3D protein structure can be determined by the values ω_i, for $i = 4, \ldots, n$.

Due to the properties of DMDGP orders, we can also know *a priori* all the values $\cos \omega_i$, for $i = 4, \ldots, n$, given by [62]

$$\cos \omega_i = \frac{2d^2_{i-2,i-1}(d^2_{i-3,i-2} + d^2_{i-2,i} - d^2_{i-3,i}) - (d_{i-3,i-2,i-1})(d_{i-2,i-1,i})}{\sqrt{4d^2_{i-3,i-2}d^2_{i-2,i-1} - (d^2_{i-3,i-2,i-1})}\sqrt{4d^2_{i-2,i-1}d^2_{i-2,i} - (d^2_{i-2,i-1,i})}}, \quad (5.3)$$

where

$$d_{i-3,i-2,i-1} = d^2_{i-3,i-2} + d^2_{i-2,i-1} - d^2_{i-3,i-1},$$
$$d_{i-2,i-1,i} = d^2_{i-2,i-1} + d^2_{i-2,i} - d^2_{i-1,i},$$

implying that the molecular structure is defined by choosing $+$ or $-$ from

$$\sin \omega_i = \pm\sqrt{1 - \cos^2 \omega_i}.$$

Based on the previous determined positions for $v_1, \ldots, v_{i-1}$, BP algorithm calculates the two possible values for $\boldsymbol{x}_i = (x_i, y_i, z_i) \in \mathbb{R}^3$, using the following expression [59]:

$$\begin{bmatrix} x_i \\ y_i \\ z_i \\ 1 \end{bmatrix} = B_1 B_2 B_3 \cdots B_i \begin{bmatrix} 0 \\ 0 \\ 0 \\ 1 \end{bmatrix}, \tag{5.4}$$

where

$$B_1 = \begin{bmatrix} 1 & 0 & 0 & 0 \\ 0 & 1 & 0 & 0 \\ 0 & 0 & 1 & 0 \\ 0 & 0 & 0 & 1 \end{bmatrix}, \quad B_2 = \begin{bmatrix} -1 & 0 & 0 & -d_2 \\ 0 & 1 & 0 & 0 \\ 0 & 0 & -1 & 0 \\ 0 & 0 & 0 & 1 \end{bmatrix},$$

$$B_3 = \begin{bmatrix} -\cos\theta_3 & -\sin\theta_3 & 0 & -d_3\cos\theta_3 \\ \sin\theta_3 & -\cos\theta_3 & 0 & d_3\sin\theta_3 \\ 0 & 0 & 1 & 0 \\ 0 & 0 & 0 & 1 \end{bmatrix},$$

and

$$B_i = \begin{bmatrix} -\cos\theta_i & -\sin\theta_i & 0 & -d_i\cos\theta_i \\ \sin\theta_i\cos\omega_i & -\cos\theta_i\cos\omega_i & -\sin\omega_i & d_i\sin\theta_i\cos\omega_i \\ \sin\theta_i\sin\omega_i & -\cos\theta_i\sin\omega_i & \cos\omega_i & d_i\sin\theta_i\sin\omega_i \\ 0 & 0 & 0 & 1 \end{bmatrix},$$

for $i = 4, \ldots, n$.

Following [62], Eq. (5.4) can be rewritten as

$$\begin{bmatrix} x_i \\ y_i \\ z_i \\ 1 \end{bmatrix} = Q_{i-1} B_i \begin{bmatrix} 0 \\ 0 \\ 0 \\ 1 \end{bmatrix}$$

$$= \begin{bmatrix} d_i\left[-q_{11}^{i-1}\cos\theta_i + \sin\theta_i\left(q_{12}^{i-1}\cos\omega_i + q_{13}^{i-1}\sin\omega_i\right)\right] + q_{14}^{i-1} \\ d_i\left[-q_{21}^{i-1}\cos\theta_i + \sin\theta_i\left(q_{22}^{i-1}\cos\omega_i + q_{23}^{i-1}\sin\omega_i\right)\right] + q_{24}^{i-1} \\ d_i\left[-q_{31}^{i-1}\cos\theta_i + \sin\theta_i\left(q_{32}^{i-1}\cos\omega_i + q_{33}^{i-1}\sin\omega_i\right)\right] + q_{34}^{i-1} \\ 1 \end{bmatrix},$$

where $Q_{i-1} = B_1 \cdots B_{i-1}$ is given by

$$Q_{i-1} = \begin{bmatrix} q_{11}^{i-1} & q_{12}^{i-1} & q_{13}^{i-1} & q_{14}^{i-1} \\ q_{21}^{i-1} & q_{22}^{i-1} & q_{23}^{i-1} & q_{24}^{i-1} \\ q_{31}^{i-1} & q_{32}^{i-1} & q_{33}^{i-1} & q_{34}^{i-1} \\ 0 & 0 & 0 & 1 \end{bmatrix}.$$

This means that to obtain the Cartesian coordinates $\boldsymbol{x}_i$, in addition to the internal coordinates d_i, θ_i, ω_i, we have to use the elements of the matrix Q_{i-1}, which depends on all the internal coordinates of the previous points. See [87] for other details.

5.4 Geometric Algebra Approach

To the best of our knowledge, the first time that a strong connection between distance geometry and geometric algebra was established in 1993, by Dress and Havel [31]. In [17, 18, 62], based on geometric algebra, an analytical expression for the Cartesian coordinates $\boldsymbol{x}_i$ was presented, using only $\boldsymbol{x}_{i-1}, \boldsymbol{x}_{i-2}, \boldsymbol{x}_{i-3}$ and the internal coordinates d_i, θ_i, ω_i, as explained below.

Using the expression (5.4), the first three atoms of the molecule can be fixed at positions

$$\boldsymbol{x}_1 = \begin{bmatrix} 0 \\ 0 \\ 0 \end{bmatrix}, \ \boldsymbol{x}_2 = \begin{bmatrix} -d_2 \\ 0 \\ 0 \end{bmatrix}, \ \boldsymbol{x}_3 = \begin{bmatrix} -d_2 + d_3 \cos\theta_3 \\ d_3 \sin\theta_3 \\ 0 \end{bmatrix}.$$

Following the notation proposed in [62], let the vector $\boldsymbol{r}_i \in \mathbb{R}^3$, for $i = 3, \ldots, n$, be defined by

$$\boldsymbol{r}_i = \boldsymbol{x}_i - \boldsymbol{x}_{i-1}.$$

For $i = 4, \ldots, n$, define the vector $\boldsymbol{r}_i^{\theta}$ and the rotor R_i^{θ} as

$$\boldsymbol{r}_i^{\theta} = -\left(\frac{d_i}{d_{i-1}}\right)\boldsymbol{r}_{i-1}$$

and

$$R_i^{\theta} = \cos\left(\frac{\theta_i}{2}\right) - \sin\left(\frac{\theta_i}{2}\right) B_i^{\theta},$$

where B_i^θ is the normalized bivector $\boldsymbol{r}_{i-2} \wedge \boldsymbol{r}_{i-1}$ and $\theta_i \in (0, \pi)$. Using $\boldsymbol{r}_i^\theta$ and R_i^θ, for $i = 4, \ldots, n$, let the vector $\boldsymbol{r}_i^\omega$ be defined by

$$\boldsymbol{r}_i^\omega = \boldsymbol{r}_i^\theta \left(R_i^\theta\right)^2.$$

For $i = 4, \ldots, n$, define the rotor R_i^ω as

$$R_i^\omega = \cos\left(\frac{\omega_i}{2}\right) - \sin\left(\frac{\omega_i}{2}\right) B_i^\omega,$$

where B_i^ω is the normalized bivector (orthogonal to $\boldsymbol{r}_{i-1}$) given by

$$B_i^\omega = \boldsymbol{r}_{i-1} \cdot e_{123},$$

where $\omega_i \in [0, 2\pi]$. Using $\boldsymbol{r}_i^\omega$ and R_i^ω, for $i = 4, \ldots, n$, we get

$$\boldsymbol{r}_i = R_i^\omega \boldsymbol{r}_i^\omega \tilde{R}_i^\omega,$$

which implies that

$$\boldsymbol{x}_i = \boldsymbol{x}_{i-1} - ABC,$$

where

$$A = \left(\cos\left(\frac{\omega_i}{2}\right) - \frac{\boldsymbol{r}_{i-1} e_{123}}{||\boldsymbol{r}_{i-1}||} \sin\left(\frac{\omega_i}{2}\right)\right) \left(\frac{d_i}{d_{i-1}} \boldsymbol{r}_{i-1}\right),$$

$$B = \left(\cos\left(\frac{\theta_i}{2}\right) - \frac{\boldsymbol{r}_{i-2} \wedge \boldsymbol{r}_{i-1}}{||\boldsymbol{r}_{i-2} \wedge \boldsymbol{r}_{i-1}||} \sin\left(\frac{\theta_i}{2}\right)\right)^2,$$

$$C = \left(\cos\left(\frac{\omega_i}{2}\right) - \frac{\boldsymbol{r}_{i-1} e_{123}}{||\boldsymbol{r}_{i-1}||} \sin\left(\frac{\omega_i}{2}\right)\right).$$

After some algebraic manipulations [62], we can eliminate the exterior and geometric products to obtain

$$\boldsymbol{x}_i = -\left(\frac{d_i}{||\boldsymbol{x}_{i-1} - \boldsymbol{x}_{i-2}||}\right) (\alpha + \beta (\gamma + \delta)), \tag{5.5}$$

where $\beta \in \mathbb{R}$ and $\alpha, \gamma, \delta \in \mathbb{R}^3$, given by

$$\alpha = \left(\cos(\theta_i) - \frac{||\boldsymbol{x}_{i-1} - \boldsymbol{x}_{i-2}||}{d_i}\right) \boldsymbol{x}_{i-1} - \cos(\theta_i) \boldsymbol{x}_{i-2},$$

$$\beta = \frac{\sin(\theta_i)}{||(\boldsymbol{x}_{i-1} - \boldsymbol{x}_{i-2}) \times (\boldsymbol{x}_{i-2} - \boldsymbol{x}_{i-3})||},$$

$$\gamma = \cos(\omega_i)\Big(||\boldsymbol{x}_{i-1} - \boldsymbol{x}_{i-2}||^2(\boldsymbol{x}_{i-2} - \boldsymbol{x}_{i-3}) - (\boldsymbol{x}_{i-1} - \boldsymbol{x}_{i-2}) \cdot$$
$$(\boldsymbol{x}_{i-2} - \boldsymbol{x}_{i-3})(\boldsymbol{x}_{i-1} - \boldsymbol{x}_{i-2})\Big),$$
$$\delta = \sin(\omega_i)||\boldsymbol{x}_{i-1} - \boldsymbol{x}_{i-2}||\,((\boldsymbol{x}_{i-1} - \boldsymbol{x}_{i-2}) \times (\boldsymbol{x}_{i-2} - \boldsymbol{x}_{i-3}))\,.$$

In [80], using the concept of polyspherical coordinates, the authors present a similar expression to (5.5).

5.5 Uncertainties from NMR Data

The main step of the BP algorithm is to calculate the two possible positions for vertex v_i, $i > 3$, in terms of the positions of vertices $v_{i-3}, v_{i-2}, v_{i-1}$ and the distances $d_{i-3,i}, d_{i-2,i}, d_{i-1,i}$. The distances $d_{i-1,i}$ and $d_{i-2,i}$ are related to bond lengths and bond angles of a protein, considered as precise values, but the distances $d_{i-3,i}$, in general, are provided by NMR data, implying that they may not be precise, being represented by an interval of real numbers $[\underline{d}_{i-3,i}, \overline{d}_{i-3,i}]$. Thus, it is necessary to solve the system

$$\begin{aligned} \underline{d}_{i-3,i} \le ||\boldsymbol{x}_i - \boldsymbol{x}_{i-3}|| &\le \overline{d}_{i-3,i}, \\ ||\boldsymbol{x}_i - \boldsymbol{x}_{i-2}|| &= d_{i-2,i}, \\ ||\boldsymbol{x}_i - \boldsymbol{x}_{i-1}|| &= d_{i-1,i}, \end{aligned} \tag{5.6}$$

where $d_{i-3,i} \in [\underline{d}_{i-3,i}, \overline{d}_{i-3,i}]$.

From the expression (5.3), for each interval related to the distance $d_{i-3,i}$, we can obtain an associated interval for $\cos\omega_i$. Using expression (5.5), we can parameterize the position of vertex v_i, for $i = 4, \ldots, n$, as

$$\boldsymbol{x}_i(\cos\omega_i) = \boldsymbol{p}_1 + (\cos\omega_i)\,\boldsymbol{p}_2 \pm \sqrt{1 - (\cos\omega_i)^2}\,\boldsymbol{p}_3, \tag{5.7}$$

where $\boldsymbol{p}_1, \boldsymbol{p}_2, \boldsymbol{p}_3 \in \mathbb{R}^3$ are given in terms of d_i, θ_i and $\boldsymbol{x}_{i-1}, \boldsymbol{x}_{i-2}, \boldsymbol{x}_{i-3}$:

$$\boldsymbol{p}_1 = -\left(\frac{d_i}{||\boldsymbol{x}_{i-1} - \boldsymbol{x}_{i-2}||}\right)\left(\left(\cos\theta_i - \frac{||\boldsymbol{x}_{i-1} - \boldsymbol{x}_{i-2}||}{d_i}\right)\boldsymbol{x}_{i-1} - (\cos\theta_i)\,\boldsymbol{x}_{i-2}\right),$$

$$\boldsymbol{p}_2 = -\left(\frac{d_i}{||\boldsymbol{x}_{i-1} - \boldsymbol{x}_{i-2}||}\right)\left(\frac{\sin\theta_i}{||(\boldsymbol{x}_{i-1} - \boldsymbol{x}_{i-2}) \times (\boldsymbol{x}_{i-2} - \boldsymbol{x}_{i-3})||}\right)$$
$$\Big(||\boldsymbol{x}_{i-1}-\boldsymbol{x}_{i-2}||^2(\boldsymbol{x}_{i-2}-\boldsymbol{x}_{i-3})-(\boldsymbol{x}_{i-1}-\boldsymbol{x}_{i-2}) \cdot (\boldsymbol{x}_{i-2}-\boldsymbol{x}_{i-3})(\boldsymbol{x}_{i-1}-\boldsymbol{x}_{i-2})\Big),$$

$$\boldsymbol{p}_3 = -\left(\frac{d_i}{||\boldsymbol{x}_{i-1} - \boldsymbol{x}_{i-2}||}\right)\left(\frac{\sin\theta_i}{||(\boldsymbol{x}_{i-1} - \boldsymbol{x}_{i-2}) \times (\boldsymbol{x}_{i-2} - \boldsymbol{x}_{i-3})||}\right)$$
$$||\boldsymbol{x}_{i-1} - \boldsymbol{x}_{i-2}||\,((\boldsymbol{x}_{i-1} - \boldsymbol{x}_{i-2}) \times (\boldsymbol{x}_{i-2} - \boldsymbol{x}_{i-3}))\,.$$

Since d_i, θ_i are given as part of the input of the problem and $\boldsymbol{x}_{i-1}, \boldsymbol{x}_{i-2}, \boldsymbol{x}_{i-3}$ are previously calculated, the only variable is $\cos \omega_i \in [-1, 1]$.

As illustrated in [62], the analytical expression (5.7) for the position of vertex v_i in terms of the positions of the three previous ones and the corresponding distances can be useful in the BP algorithm, but the positions for $v_{i-3}, v_{i-2}, v_{i-1}$ are assumed to be fixed.

Computational results presented in [59] demonstrate that the use of the expressions (5.4), instead of resolutions of quadratic systems, is more stable in BP algorithm. However, none of these approaches deals well with interval distances. Some preliminary results for the application of the homogeneous matrices in the context of interval distances are given in [21].

5.6 Conformal Geometric Algebra Approach

Geometrically, property 2 of the DMDGP (see Sect. 5.2) means that, at each iteration of the BP algorithm, we have to intersect three spheres centered at the positions for vertices $v_{i-3}, v_{i-2}, v_{i-1}$ with radius given by $d_{i-3,i}, d_{i-2,i}, d_{i-1,i}$, respectively. Since the centers are not collinear [see Eq. (5.2)], the sphere intersection will provide up to two possible positions for v_i. To avoid solutions modulo rotations and translations (property 1 of the DMDGP), the first three vertices are fixed and, for vertex v_i, $i > 3$, an additional distance $d_{j,i}$ $(j < i - 3)$ implies that we have to intersect four spheres. If the centers of these four spheres are not in the same plane, we obtain only one possible position $\boldsymbol{x}_i$ for v_i (if $||\boldsymbol{x}_i - \boldsymbol{x}_j|| = d_{j,i}$) or an empty set. In this case, we have to consider the second possible position for v_{i-1} and repeat the same strategy. If the DMDGP solution set is not empty, this procedure is finite and ends when a position for vertex v_n is found such that all positions $\boldsymbol{x}_1, \ldots, \boldsymbol{x}_n \in \mathbb{R}^3$ satisfy the DMDGP equations [59].

Due to NMR uncertainties, we already mentioned that the distances $d_{i-3,i}$ may be represented by interval distances $[\underline{d}_{i-3,i}, \overline{d}_{i-3,i}]$. In [61], an extension, called *i*BP, of the BP algorithm was proposed to deal with interval distances, where the idea is to sample distance values from the intervals $[\underline{d}_{i-3,i}, \overline{d}_{i-3,i}]$. This is done because it is difficult to do sphere intersection using linear algebra, when the data are not precise. In fact, the problem now is much more complicated, because uncertainties in distance values imply also uncertainties in the centers of the spheres, not only in their radii.

The main problem of sampling distances from $[\underline{d}_{i-3,i}, \overline{d}_{i-3,i}]$ is that if we choose many values, the search space increases exponentially, and for small samples, no solution is found [1, 14, 89].

Using conformal geometric algebra, we avoid the sampling process and can calculate intersection of spheres considering the uncertainties associated to their centers and radius.

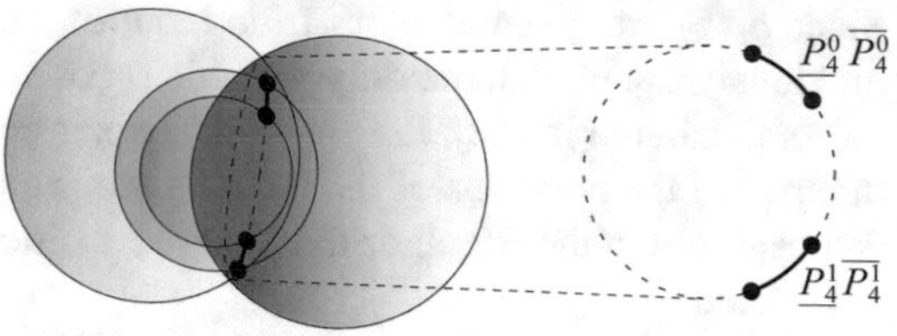

Fig. 5.2 Intersection between two spheres and one spherical shell

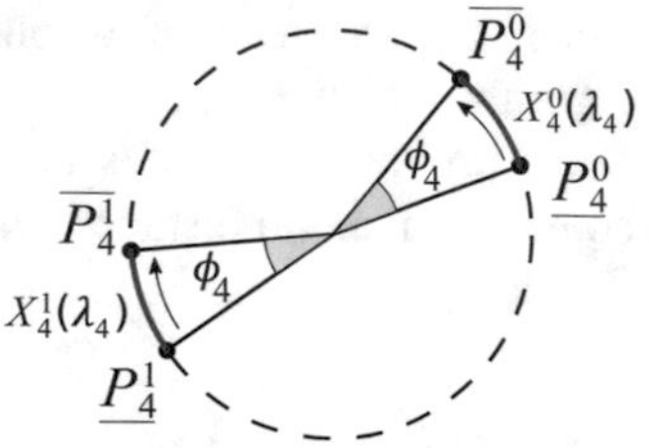

Fig. 5.3 Arcs with their orientations

Sphere Intersection with Uncertainties

We will follow the arguments presented in [2, 3] to explain how conformal geometric algebra can be used to model uncertainties in the DMDGP. First, notice that when $d_{i-3,i}$ is an interval distance, we have to intersect two spheres with one spherical shell resulting in two arcs, instead of two points in $\mathbb{R}^3$ (Fig. 5.2).

Considering that the distance $d_{1,4}$ is represented by the interval $[\underline{d}_{1,4}, \overline{d}_{1,4}]$, we first obtain the points from the intersection of the spheres centered at the positions for v_1, v_2, v_3 with radius $\underline{d}_{1,4}, d_{2,4}, d_{3,4}$, resulting in $\underline{P_4^0}$ and $\underline{P_4^1}$, and with radius $\overline{d}_{1,4}, d_{2,4}, d_{3,4}$, resulting in $\overline{P_4^0}$ and $\overline{P_4^1}$ (Fig. 5.2). The points from the sphere intersection can be calculated in the classical way or by extracting them from the point pairs [2] generated from $\underline{S}_{1,4} \wedge S_{2,4} \wedge S_{3,4}$ and $\overline{S}_{1,4} \wedge S_{2,4} \wedge S_{3,4}$ (in the conformal model, we denote $S_{i,j}$ by the sphere centered at the position of vertex v_i, denoted by X_i, with radius $d_{i,j}$), where underline and overline indicate the use of $\underline{d}_{1,4}$ and $\overline{d}_{1,4}$, respectively.

With the starting and the ending points of an arc, we can define a rotor acting on that. In conformal geometric algebra, a rotor can be defined by its rotation axis and rotation angle. For v_4, the rotation axis is given by X_2 and X_3, denoted by z_4, and the rotation angle ϕ_4 (in radians) is the angle corresponding to the arcs $\underline{P_4^0}\,\overline{P_4^0}$ and $\underline{P_4^1}\,\overline{P_4^1}$ (Fig. 5.3).

Defining the rotor R_4 by

$$R_4 = \cos\left(\frac{\lambda_4}{2}\right) - \sin\left(\frac{\lambda_4}{2}\right) z_4^*, \quad 0 \leq \lambda_4 \leq \phi_4,$$

where $z_4 = X_2 \wedge X_3 \wedge e_\infty$, the two possible arcs are the following (z_4^* is the dual of z_4 and $\tilde{R}$ is the reverse of R):

$$X_4^0(\lambda_4) = R_4 \underline{P_4^0} \tilde{R}_4,$$

$$X_4^1(-\lambda_4) = R_4 \underline{P_4^1} \tilde{R}_4.$$

To simplify the notation, we will also use the symbols $\underline{P_4^0}$ and $\underline{P_4^1}$ in the conformal model, *i.e.* $X_4^0(0) = \underline{P_4^0}$ and $X_4^1(0) = \underline{P_4^1}$. Negative values in $X_4^1(-\lambda_4)$ are necessary to invert the orientation used in $\overline{\underline{P_4^0}P_4^0}$ (Fig. 5.3).

These results can be easily generalized for any $i > 4$, given by

$$X_i^0(\lambda_i) = R_i \underline{P_i^0} \tilde{R}_i,$$

$$X_i^1(-\lambda_i) = R_i \underline{P_i^1} \tilde{R}_i,$$

where

$$R_i = \cos\left(\frac{\lambda_i}{2}\right) - \sin\left(\frac{\lambda_i}{2}\right) z_i^*, \quad 0 \le \lambda_i \le \phi_i,$$

and

$$z_i = X_{i-2} \wedge X_{i-1} \wedge e_\infty.$$

$X_i^0(\lambda_i)$ and $X_i^1(-\lambda_i)$ are the analytical solutions to the system (5.6), which generalize the case for a precise value of $d_{i-3,i}$ ($\lambda_i = 0$). All the values ϕ_i, for $i > 3$, can be computed *a priori* based on the DMDGP definition and the values for intervals $[\underline{d}_{i-3,i}, \overline{d}_{i-3,i}]$ (notice that sample values from $[\underline{d}_{i-3,i}, \overline{d}_{i-3,i}]$ imply sample values in $[0, \phi_i]$).

The expressions for $X_i^0(\lambda_i)$ and $X_i^1(-\lambda_i)$ describe all the points in the related arcs, considering fixed points (X_{i-2} and X_{i-1}) for the rotation axis z_i. However, in order to avoid sampling process, we have to consider the effect of changing points in the arcs. Using conformal geometric algebra, we can do that and continue the search without sampling process.

Before analyzing what happens with the position of vertex v_5, notice that the circle C_4 that contains the arcs $\overline{\underline{P_4^0}P_4^0}$ and $\overline{\underline{P_4^1}P_4^1}$ is given by

$$C_4 = S_{2,4} \wedge S_{3,4},$$

whose points $X_4(\lambda_4)$ are generated by the rotor

$$R_4 = \cos\left(\tfrac{\lambda_4}{2}\right) - \sin\left(\tfrac{\lambda_4}{2}\right) z_4^*, \quad 0 \le \lambda_4 \le 2\pi,$$

where

$$z_4 = X_2 \wedge X_3 \wedge e_\infty.$$

Since X_2 and X_3 are fixed, λ_4 is the only parameter involved.

For v_5, we have to consider the three predecessors v_2, v_3, v_4. Now, the center of the sphere $S_{4,5}$ is not fixed, which implies that the circle $C_5 = S_{3,5} \wedge S_{4,5}$ also moves. The rotor R_5 is given by

$$R_5 = \cos\left(\tfrac{\lambda_5}{2}\right) - \sin\left(\tfrac{\lambda_5}{2}\right) z_5^*, \quad 0 \leq \lambda_5 \leq 2\pi,$$

where

$$z_5 = X_3 \wedge X_4(\lambda_4) \wedge e_\infty.$$

From the expression for z_5, we can see that the rotation axis for R_5 also changes when λ_4 varies. The points in the circle C_4 can be described by

$$X_4(\lambda_4) = R_4 \underline{P_4^0} \tilde{R}_4,$$

and for $\lambda_4 = 0$, we have $X_4(0) = \underline{P_4^0}$ and $z_5 = X_3 \wedge \underline{P_4^0} \wedge e_\infty$. Doing some calculations, we obtain

$$z_5 = R_4(X_3 \wedge \underline{P_4^0} \wedge e_\infty)\tilde{R}_4,$$

which means that we can fix one point in the circle C_4 and "transfer" the movement of the circle C_5 to the rotor R_4. An important consequence is that

$$X_5(\lambda_4, \lambda_5) = R_5 R_4 \underline{P_5^0} \tilde{R}_4 \tilde{R}_5,$$

implying that we can describe the whole set of possible positions for v_5, without sampling values from the interval $[\underline{d}_{1,4}, \overline{d}_{1,4}]$. The position X_5 depends on the "local action" of R_5, through the axis determined by the "global action" of R_4.

By induction [3], we can prove that, for $i = 4, \ldots, n$,

$$X_i(\lambda_4, \ldots, \lambda_i) = (R_i \cdots R_4)\, \underline{P_i^0} \left(\tilde{R}_4 \cdots \tilde{R}_i\right),$$

where

$$R_i = \cos\left(\frac{\lambda_i}{2}\right) - \sin\left(\frac{\lambda_i}{2}\right) z_i^*, \quad 0 \leq \lambda_i \leq 2\pi,$$

$$z_i = (R_i \cdots R_4)\left(\underline{P_{i-2}^0} \wedge \underline{P_{i-1}^0} \wedge e_\infty\right)\left(\tilde{R}_4 \cdots \tilde{R}_i\right),$$

and $\underline{P_i^0}$ is one of the points obtained from the intersection $\underline{S}_{i-3,i} \wedge S_{i-2,i} \wedge S_{i-1,i}$. The interval distances $[\underline{d}_{i-3,i}, \overline{d}_{i-3,i}]$ reduce the intervals related to λ_i, i.e. $\lambda_i \in [0, \phi_i]$.

Example

Let us consider a small example (the same provided in [3]) just to illustrate the difference between the classical approach and the one based on conformal geometric algebra. We want to solve a DMDGP instance with the vertex order $v_1, v_2, v_3, v_4, v_5, v_6$ and the following associated distances:

$$
\begin{aligned}
d_{i-1,i} &= 1,\, i = 2, \ldots, 6, \\
d_{i-2,i} &= \sqrt{3},\, i = 3, \ldots, 6, \\
d_{1,4} &= 2.15, \\
d_{2,5} &\in [2.20, 2.60], \\
d_{3,6} &\in [2.40, 2.60], \\
d_{1,5} &\in [2.45, 2.55].
\end{aligned}
$$

First, we can fix the positions for v_1, v_2, v_3, given by

$$
\boldsymbol{x}_1 = \begin{bmatrix} 0 \\ 0 \\ 0 \end{bmatrix}, \boldsymbol{x}_2 = \begin{bmatrix} -1 \\ 0 \\ 0 \end{bmatrix}, \boldsymbol{x}_3 = \begin{bmatrix} -1.5 \\ \frac{\sqrt{3}}{2} \\ 0 \end{bmatrix}.
$$

The distance $d_{1,4}$ is a precise value, which implies that we have two possible positions (instead of two arcs) for v_4. Choosing one of them, since both are feasible, we get

$$
\boldsymbol{x}_4 = \begin{bmatrix} -1.311 \\ 1.552 \\ 0.702 \end{bmatrix}.
$$

For vertex v_5, we have two possible arcs, since $d_{2,5}$ is an interval distance. Using the classical approach, we have to sample values from the interval $[\underline{d}_{2,5}, \overline{d}_{2,5}] = [2.20, 2.60]$ to solve the system

$$
\begin{aligned}
\underline{d}_{2,5} \le ||\boldsymbol{x}_5 - \boldsymbol{x}_2|| &\le \overline{d}_{2,5}, \\
||\boldsymbol{x}_5 - \boldsymbol{x}_3|| &= d_{3,5}, \\
||\boldsymbol{x}_5 - \boldsymbol{x}_4|| &= d_{4,5}.
\end{aligned}
$$

Since the distance $d_{1,5}$ is available, it must be used to validate the candidates for positions of v_5. For example, selecting the values $\{2.25, 2.35, 2.45, 2.55\}$, no solution is found. It is necessary a larger sample, given by

$$
\{2.225, 2.275, 2.325, 2.375, 2.425, 2.475, 2.525, 2.575\},
$$

to find a solution for v_5 that satisfies

$$\underline{d}_{1,5} \leq ||\boldsymbol{x}_5 - \boldsymbol{x}_1|| \leq \overline{d}_{1,5},$$

given by

$$\boldsymbol{x}_5 = \begin{bmatrix} -0.779 \\ 2.368 \\ 0.474 \end{bmatrix}.$$

For the next vertex v_6, any value from the interval $[\underline{d}_{3,6}, \overline{d}_{3,6}]$ can be selected, since no distance $d_{i,6}$, $i < 3$, is available.

Because of the distance $d_{1,5}$, in order to continue the search from v_5 to v_6 it was necessary to improve the sample from $[\underline{d}_{2,5}, \overline{d}_{2,5}]$ until a solution is found. If we have a little bigger instance such that only at vertex v_{10}, for example, an additional distance $d_{i,10}$, $i < 7$, is available, we do not know (during the calculations for v_5) how refined the sample from $[\underline{d}_{2,5}, \overline{d}_{2,5}]$ must be in order to obtain a position for v_{10} that satisfies

$$\underline{d}_{i,10} \leq ||\boldsymbol{x}_{10} - \boldsymbol{x}_i|| \leq \overline{d}_{i,10}.$$

Using conformal geometric algebra, we first consider the conformal representations of the points $\boldsymbol{x}_1, \boldsymbol{x}_2, \boldsymbol{x}_3, \boldsymbol{x}_4$, given by

$$\begin{aligned} X_1 &= e_0, \\ X_2 &= e_0 + \boldsymbol{x}_2 + \frac{1}{2}||\boldsymbol{x}_2||^2 e_\infty, \\ X_3 &= e_0 + \boldsymbol{x}_3 + \frac{1}{2}||\boldsymbol{x}_3||^2 e_\infty, \\ X_4 &= e_0 + \boldsymbol{x}_4 + \frac{1}{2}||\boldsymbol{x}_4||^2 e_\infty, \end{aligned}$$

and the conformal representations of the points obtained from the intersection of the spheres centered at $\boldsymbol{x}_2, \boldsymbol{x}_3, \boldsymbol{x}_4$ with radius $\underline{d}_{2,5}, d_{3,5}, d_{4,5}$, given by

$$\begin{aligned} \underline{P_5^0} &= e_0 - 0.409e_1 + 1.981e_2 + 0.753e_3 + 2.329e_\infty, \\ \underline{P_5^1} &= e_0 - 1.502e_1 + 1.350e_2 + 1.663e_3 + 3.422e_\infty. \end{aligned}$$

Points $\overline{P_5^0}$ and $\overline{P_5^1}$ are obtained just replacing $\underline{d}_{2,5}$ by $\overline{d}_{2,5}$.

Defining the rotor R_5 to act on the arc $\underline{P_5^0}\,\overline{P_5^0}$, we get

$$R_5 = \cos\left(\tfrac{\lambda_5}{2}\right) - \sin\left(\tfrac{\lambda_5}{2}\right) z_5^*, \; 0 \leq \lambda_5 \leq \phi_5,$$

where $z_5 = X_3 \wedge X_4 \wedge e_\infty$ and $\phi_5 = 1.453$ is the angle corresponding to the arc $\overline{\underline{P_5^0} P_5^0}$, whose points are given by

$$X_5^0(\lambda_5) = R_5 \underline{P_5^0} \tilde{R}_5.$$

The same procedure can be used to obtain the points in the arc $\overline{\underline{P_5^1} P_5^1}$.

Intersecting $S_{3,5} \wedge S_{4,5}$ with the spherical shell defined by the interval distance $d_{1,5}$, we can eliminate the arc $\overline{\underline{P_5^1} P_5^1}$ (since it does not satisfy $d_{1,5}$) and, at the same time, reduce to [0.559, 0.734] the interval related to λ_5.

For the next vertex v_6, we intersect the spheres centered at $X_3, X_4, \underline{P_5^0}$ with radius $\underline{d}_{3,6}, d_{4,6}, d_{5,6}$, respectively, resulting in (using $\overline{d}_{3,6}$, $\overline{P}_6^0$ and $\overline{P}_6^1$ are obtained in a similar way)

$$\underline{P_6^0} = e_0 - 0.2657e_1 + 2.892e_2 + 0.3658e_3 + 4.283e_\infty,$$

$$\underline{P_6^1} = e_0 + 0.2584e_1 + 1.662e_2 + 1.426e_3 + 2.432e_\infty,$$

and define the rotor R_6 to act on the arc $\overline{\underline{P_6^0} P_6^0}$, given by

$$R_6 = \cos\left(\frac{\lambda_6}{2}\right) - \sin\left(\frac{\lambda_6}{2}\right) z_6^*, \quad 0 \leq \lambda_6 \leq \phi_6,$$

where $z_6 = R_5 \left(X_4 \wedge \underline{P_5^0} \wedge e_\infty\right) \tilde{R}_5$ and $\phi_6 = 0.823$. The points in the arc $\overline{\underline{P_6^0} P_6^0}$ are given by

$$X_6^0(\lambda_5, \lambda_6) = R_6 R_5 \underline{P_6^0} \tilde{R}_5 \tilde{R}_6.$$

Since there is no distance $d_{i,6}$, $i < 3$, the solutions to the problem are given by

$$X_1, X_2, X_3, X_4, X_5^0(\lambda_{5,}), X_6^0(\lambda_{5,}\lambda_6),$$

and

$$X_1, X_2, X_3, X_4, X_5^0(\lambda_{5,}), X_6^1(\lambda_{5,} - \lambda_6),$$

where $\lambda_5 \in [0.5595, 0.734]$ and $\lambda_6 \in [0, 0.823]$.

5.7 Exercises

E.5.1 In Sect. 5.3, it was mentioned that, for some $i > 3$, when there is an "extra" edge $\{v_j, v_i\} \in E$, $j < i - 3$, in addition to the required edges in the DMDGP definition ($\{v_{i-3}, v_i\}, \{v_{i-2}, v_i\}, \{v_{i-1}, v_i\} \in E$), the position for vertex v_i can be obtained from a linear system in the variable $\boldsymbol{x}_i$. Define this linear system and find a condition to guarantee a unique solution.

E.5.2 Describe the main steps of the BP algorithm, also mentioned in Sect. 5.3.

E.5.3 Explain how expression (5.7) can be obtained from expression (5.5).

E.5.4 Explain how the angles ϕ_i, for $i > 3$ (defined in Sect. 5.6), can be computed *a priori* based on the DMDGP definition and the values for intervals $[\underline{d}_{i-3,i}, \overline{d}_{i-3,i}]$.

E.5.5 Based on the example given in Sect. 5.6, what are the main differences between the classical and the conformal geometric algebra approaches? Can you calculate all the values involved in the example?

References

1. A. Agra, R. Figueiredo, C. Lavor, N. Maculan, A. Pereira, C. Requejo, Feasibility check for the distance geometry problem: an application to molecular conformations. Int. Trans. Oper. **24**(5), 1023–1040 (2017)
2. R. Alves, C. Lavor, Geometric algebra to model uncertainties in the discretizable molecular distance geometry problem. Adv. Appl. Clifford Algebr. **27**(1), 439–452 (2017)
3. R. Alves, C. Lavor, C. Souza, M. Souza, Clifford algebra and discretizable distance geometry. Math. Methods Appl. Sci. **41**, 3999–4346 (2018)
4. E. Artin, *Geometric Algebra*. Tracts in Pure and Applied Mathematics, vol. 3 (Interscience, New York, 1957)
5. E. Bayro-Corrochano, D. Kähler, Motor algebra approach for computing the kinematics of robot manipulators. J. Robot. Syst. **17**(9), 495–516 (2000)
6. E. Bayro-Corrochano, D. Kähler, Kinematics of robot manipulators in the motor algebra, in *Geometric Computing with Clifford Algebras: Theoretical Foundations and Applications in Computer Vision and Robotics*, ed. by G. Sommer (Springer, London, 2001), pp. 471–488
7. E. Bayro-Corrochano, J. Zamora-Esquivel, Differential and inverse kinematics of robot devices using conformal geometric algebra. Robotica **25**(1), 43–61 (2007)
8. S. Billinge, P. Duxbury, D. Gonçalves, C. Lavor, A. Mucherino, Assigned and unassigned distance geometry: applications to biological molecules and nanostructures. 4OR **14**(4), 337–376 (2016)
9. A. Brünger, M. Nilges, Computational challenges for macromolecular structure determination by X-ray crystallography and solution NMR-spectroscopy. Q. Rev. Biophys. **26**(1), 49–125 (1993)
10. S.R. Buss, Introduction to inverse kinematics with Jacobian transpose, pseudoinverse and Damped Least Squares methods. Technical report, University of California, San Diego (2009)
11. S.R. Buss, J.S. Kim, Selectively Damped Least Squares for inverse kinematics. J. Graph. Tools **10**(3), 37–49 (2005)
12. J. Cameron, J. Lasenby, Oriented conformal geometric algebra. Adv. Appl. Clifford Algebr. **18**(3–4), 523–538 (2008)
13. G. Casanova, *L'algèbre vectorielle*. Number 1657 in Que sais-je? Presses Universitaires de France (1976)
14. A. Cassioli, B. Bordeaux, G. Bouvier, M. Mucherino, R. Alves, L. Liberti, M. Nilges, C. Lavor, T. Malliavin, An algorithm to enumerate all possible protein conformations verifying a set of distance constraints. BMC Bioinf. **16**, 16–23 (2015)

C. Lavor et al., *A Geometric Algebra Invitation to Space-Time Physics, Robotics and Molecular Geometry*, SpringerBriefs in Mathematics,
https://doi.org/10.1007/978-3-319-90665-2

15. A. Cassioli, O. Gunluk, C. Lavor, L. Liberti, Discretization vertex orders in distance geometry. Discrete Appl. Math. **197**, 27–41 (2015)
16. P. Chang, A closed-form solution for inverse kinematics of robot manipulators with redundancy. IEEE J. Robot. Autom. **3**(5), 393–403 (1987)
17. P. Chys, Application of geometric algebra for the description of polymer conformations. J. Chem. Phys. **128**(10), 104107(1)–104107(12) (2008)
18. P. Chys, P. Chacón, Spinor product computations for protein conformations. J. Comput. Chem. **33**(21), 1717–1729 (2012)
19. W.K. Clifford, Applications of Grassmann's extensive algebra, in *Proceedings of the London Mathematical Society* (1878)
20. P. Colapinto, Articulating space: geometric algebra for parametric design – symmetry, kinematics, and curvature. PhD thesis, Media Arts and Technology Program, University of California Santa Barbara (2016)
21. T. Costa, H. Bouwmeester, W. Lodwick, C. Lavor, Calculating the possible conformations arising from uncertainty in the molecular distance geometry problem using constraint interval analysis. Inf. Sci. **415–416**, 41–52 (2017)
22. H.S.M. Coxeter, S.L. Greitzer, *Geometry Revisited* (The Mathematical Association of America, Washington, DC, 1967)
23. G. Crippen, T. Havel, *Distance Geometry and Molecular Conformation* (Wiley, New York, 1988)
24. A. Das, *The Special Theory of Relativity. A Mathematical Exposition*. Universitext (Springer, New York, 1996). Corrected second printing of the 1993 first edition
25. A. De Luca, G. Oriolo, The reduced gradient method for solving redundancy in robot arms. Robotersysteme **7**, 117–122 (1991)
26. J. Denavit, R.S. Hartenberg, A kinematic notation for lower-pair mechanisms based on matrices. J. Appl. Mech. **22**(2), 215–221 (1965)
27. P.A.M. Dirac, The quantum theory of the electron, I, II. Proc. R. Soc. Lond. **A117**, 610–624; **A118**, 351–361 (1928)
28. B. Donald, *Algorithms in Structural Molecular Biology* (MIT Press, Cambridge, 2011)
29. C. Doran, A. Lasenby, *Geometric Algebra for Physicists* (Cambridge University Press, Cambridge, 2003)
30. L. Dorst, D. Fontijne, S. Mann, *Geometric Algebra for Computer Science: An Object-Oriented Approach to Geometry* (Morgan Kaufmann, San Francisco, 2007)
31. A. Dress, T. Havel, Distance geometry and geometric algebra. Found. Phys. **23**(10), 1357–1374 (1993)
32. D.A. Drexler, Solution of the closed-loop inverse kinematics algorithm using the Crank-Nicolson method, in *IEEE 14th International Symposium on Applied Machine Intelligence and Informatics (SAMI)*, Herlany, 21–23 January 2016, pp. 351–356
33. A. Einstein, *On the Electrodynamics of Moving Bodies* (Dover, New York, 1952), pp. 37–65. Translation of "Zur Elektrodynamik bewegter Körper", *Annalen der Physik*, 17 (1905)
34. A. Fratu, L. Vermeiren, A. Dequidt, Using the redundant inverse kinematics system for collision avoidance, in *International Symposium on Electrical and Electronics Engineering (ISEEE)*, Galati, Romania, 16–18 September 2010, pp. 88–93
35. D. Gonçalves, A. Mucherino, Discretization orders and efficient computation of Cartesian coordinates for distance geometry. Optim. Lett. **8**(7), 2111–2125 (2014)
36. D. Gottlieb, Robots and topology, in *IEEE International Conference on Robotics and Automation (ICRA)*, San Francisco, CA, vol. 3, 7–10 April 1986, pp. 1689–1691
37. H.G. Grassmann, *Die lineale Ausdehnungslehre, ein neuer Zweig der Mathematik* (Otto Wiegand, Leipzig, 1844)
38. H.G. Grassmann, *Die Ausdehnungslehre. Vollständig und in strenger Form* (Adolf Enslin, Berlin, 1862)
39. H.G. Grassmann, *Extension Theory* (American Mathematical Society, Providence, 2000). Traslated from the German version *Die Ausdehnungslehre von 1862* by Lloys C. Kannenberg

40. D. Hestenes, *Space-Time Algebra* (Gordon& Breach, New York, 1966). 2nd edition: Birkhäuser 2015, with a Foreword by A. Lasenby and new "Preface after fifty years" by the author
41. D. Hestenes, Real Dirac theory, in *The Theory of the Electron*, ed. by J. Keller, Z. Oziewicz (UNAM, México, 1986), pp. 1–50
42. D. Hestenes, Spinor particle mechanics, in *Proceedings of the Fourth International Conference on Clifford Algebras and Their Applications to Mathematical Physics* (Reidel, Aachen, 1986), pp. 129–143
43. D. Hestenes, A unified language for mathematics and physics, in *Clifford Algebras and Their Applications in Mathematical Physics*, ed. by J.S.R. Chisholm, A.K. Commons (Reidel, Dordrecht/Boston, 1986), pp. 1–23
44. D. Hestenes, Mysteries and insights of Dirac theory. Annales de la Fondation Louis de Broglie **28**(3), 390–408 (2003)
45. D. Hestenes, Oersted medal lecture 2002: reforming the mathematical language of physics. Am. J. Phys. **71**(2), 104–121 (2003)
46. D. Hestenes, Spacetime physics with geometric algebra. Am. J. Phys. **71**(7), 691–714 (2003)
47. D. Hestenes. (1) Maxwell-Dirac electron theory. (2) Quantum Mechanics of the electron particle-clock. Preprints received on 22 April, 2018.
48. D. Hestenes, The genesis of geometric algebra: a personal retrospective. Adv. Appl. Clifford Algebr. **27**(1), 351–379 (2017). Opening paper of the AGACSE 2015 Proceedings
49. D. Hildenbrand, *Foundations of Geometric Algebra Computing* (Springer, New York, 2012)
50. J.M. Hollerbach, Optimum kinematic design for a seven degree of freedom manipulator, in *Robotics Research: The Second International Symposium* (MIT Press, Cambridge, 1985), pp. 215–222
51. L.S. Huang, R.K. Jiang, A new method of inverse kinematics solution for industrial 7DoF robot, in *32nd Chinese Control Conference (CCC)*, Xi'an, 26–28 July 2013, pp. 6063–6065
52. O. Ivlev, A. Gräser, An analytical method for the inverse kinematics of redundant robots, in *International Conference Advanced Robotics, Intelligent Automation and Active Systems*, Bremen, 15–17 September 1997, pp. 416–421
53. F. John, *Partial Differential Equations*, 3rd edn. Applied Mathematical Sciences, vol. 1 (Springer, New York, 1978)
54. K. Kanatani, *Understanding Geometric Algebra. Hamilton, Grassmann, and Clifford for Computer Vision and Graphics* (CRC Press, Boca Raton, 2015)
55. A. Lasenby, J. Lasenby, R. Wareham, A convariant approach to geometry using geometric algebra. Technical report, Department of Engineering, University of Cambridge (2004)
56. C. Lavor, Analytic evaluation of the gradient and Hessian of molecular potential energy functions. Phys. D Nonlinear Phenom. **227**(2), 135–141 (2007)
57. C. Lavor, N. Maculan, A function to test methods applied to global minimization of potential energy of molecules. Numer. Algorithms **35**(2–4), 287–300 (2004)
58. C. Lavor, L. Liberti, N. Maculan, Computational experience with the molecular distance geometry problem, in *Global Optimization*, ed. by J. Pintér, Nonconvex Optimization and Its Applications, vol. 85 (Springer, New York, 2006), pp. 213–225
59. C. Lavor, L. Liberti, N. Maculan, A. Mucherino, The discretizable molecular distance geometry problem. Comput. Optim. Appl. **52**(1), 115–146 (2012)
60. C. Lavor, L. Liberti, N. Maculan, A. Mucherino, Recent advances on the discretizable molecular distance geometry problem. Eur. J. Oper. Res. **219**(3), 698–706 (2012)
61. C. Lavor, L. Liberti, A. Mucherino, The interval BP algorithm for the discretizable molecular distance geometry problem with interval data. J. Glob. Optim. **56**(3), 855–871 (2013)
62. C. Lavor, R. Alves, W. Figueiredo, A. Petraglia, N. Maculan, Clifford algebra and the discretizable molecular distance geometry problem. Adv. Appl. Clifford Algebr. **25**(4), 925–942 (2015)
63. H. Li, *Invariant Algebras and Geometric Reasoning* (World Scientific, Singapore, 2008)

64. H. Li, D. Hestenes, A. Rockwood, A universal model for conformal geometries of Euclidean, spherical and double-hyperbolic spaces, in *Geometric Computing with Clifford Algebras* (Springer, New York, 2001), pp. 77–104
65. L. Liberti, C. Lavor, Six mathematical gems from the history of distance geometry. Int. Trans. Oper. Res. **23**(5), 897–920 (2016)
66. L. Liberti, C. Lavor, N. Maculan, A branch-and-prune algorithm for the molecular distance geometry problem. Int. Trans. Oper. Res. **15**(1), 1–17 (2008)
67. L. Liberti, C. Lavor, A. Mucherino, N. Maculan, Molecular distance geometry methods: from continuous to discrete. Int. Trans. Oper. Res. **18**(1), 33–51 (2011)
68. L. Liberti, C. Lavor, N. Maculan, A. Mucherino, Euclidean distance geometry and applications. SIAM Rev. **56**(1), 3–69 (2014)
69. L. Liberti, B. Masson, J. Lee, C. Lavor, A. Mucherino, On the number of realizations of certain Henneberg graphs arising in protein conformation. Discrete Appl. Math. **165**, 213–232 (2014)
70. J. Liouville, Extension au cas des trois dimensions de la question du tracé géographique, in G. Monge, *Applications de l'analyse à la géométrie* (Bachelier, Paris, 1850), pp. 609–617 (Note VI)
71. Y. Liu, D. Wang, J. Sun, L. Chang, C. Ma, Y. Ge, L. Gao, Geometric approach for inverse kinematics analysis of 6-DOF serial robot, in *IEEE International Conference on Information and Automation*, Lijiang, 8–10 August 2015, pp. 852–855
72. P. Lounesto, *Clifford Algebras and Spinors*, 2nd edn. LMS Lecture Note Series, vol. 286 (Cambridge University Press, Cambridge, 2001)
73. A.A. Michelson, E.W. Morley, On the relative motion of the earth and the luminiferous ether. Am. J. Sci. 3rd Ser. **34**, 333–345 (1887)
74. H. Minkowski, Space and time, in *The Principle of Relativity* (Dover, New York, 1952), pp. 75–91. Translation of the communication «Raum und Zeit»presented by the author to the 80th Convention of German Scientists and Doctors (Köln, 21 September, 1908)
75. A. Mucherino, C. Lavor, L. Liberti, The discretizable distance geometry problem. Optim. Lett. **6**(8), 1671–1686 (2012)
76. A. Mucherino, C. Lavor, L. Liberti, N. Maculan (eds.), *Distance Geometry: Theory, Methods, and Applications* (Springer, New York, 2013)
77. R.P. Paul, *Robot Manipulators: Mathematics, Programming and Control* (MIT Press, Cambridge, 1981)
78. W. Pauli, Zur Quantenmechanik des magnetischen Elektrons. Z. Phys. **42**, 601–623 (1927)
79. C. Perwass, *Geometric Algebra with Applications in Engineering*. Geometry and Computing, vol. 4 (Springer, New York, 2009)
80. J. Pesonen, O. Henriksson, Polymer conformations in internal (polyspherical) coordinates. J. Comput. Chem. **31**(9), 1873–1881 (2009)
81. D.L. Pieper, The kinematics of manipulation under computer control. PhD thesis, Stanford Artificial Intelligence Laboratory, Stanford University (1968)
82. M. Riesz, *Clifford Numbers and Spinors*. Fundamental Theories of Physics, vol. 54 (Kluwer Academic, New York, 1997). An edition by E. F. Bolinder and P. Lounesto of the memoir *Clifford Numbers and Spinors* by M. Riesz, Lecture Series No. 38, Institute for Fluid Dynamics and Applied Mathematics, University of Maryland, 1958. Includes an annex by Bolinder and an article by Lounesto of Riesz' work
83. O. Rodrigues, Des lois géométriques qui régissent les déplacements d'un système solide dans l'espace, et de la variation des coordonées provenant de ses déplacements considérés indépendamment des causes qui peuvent les produire. Journal des Mathématiques Pures et Appliquées **5**, 380–440 (1840)
84. J. Saxe, Embeddability of weighted graphs in k-space is strongly NP-hard, in *Proceedings of 17th Allerton Conference in Communications, Control and Computing* (1979), pp. 480–489
85. M. Schottenloher, *A Mathematical Introduction to Conformal Field Theory*. LNP, vol. 759 (Springer, Berlin, 1997). A much enlarged second edition appeared in 2008
86. J.M. Selig, Clifford algebra of points, lines and planes. Robotica **18**(5), 545–556 (2000)

87. C. Seok, E. Coutsias, Efficiency of rotational operators for geometric manipulation of chain molecules. Bull. Kor. Chem. Soc. **28**(10), 1705–1708 (2007)
88. B. Siciliano, L. Sciavicco, L. Villani, G. Oriolo, *Robotics: Modelling, Planning and Control* (Springer, New York, 2008)
89. M. Souza, C. Lavor, A. Muritiba, M. Maculan, Solving the molecular distance geometry problem with inaccurate distance data. BMC Bioinf. **14**(9), S7(1)–S7(6) (2013)
90. M.W. Spong, S. Hutchinson, M. Vidyasagar, *Robot Modeling and Control* (Wiley, New York, 2006)
91. N. Vahrenkamp, D. Muth, P. Kaiser, T. Asfour, IK-Map: an enhanced workspace representation to support inverse kinematics solvers, in *IEEE-RAS 15th International Conference on Humanoid Robots (Humanoids)*, Seoul, 3–5 November 2015, pp. 785–790
92. J. Vaz Jr., R. da Rocha Jr., *An Introduction to Clifford Algebras and Spinors* (Oxford University Press, Oxford, 2016)
93. Wikipedia, Tests of special relativity. https://en.wikipedia.org/wiki/Tests_of_special_relativity. Consulted on December 2017
94. K. Wütrich, Protein structure determination in solution by nuclear magnetic resonance spectroscopy. Science **243**, 45–50 (1989)
95. S. Xambó-Descamps, *Álgebra lineal y geometrías lineales, II* (EUNIBAR, Barcelona, 1978)
96. S. Xambó-Descamps, Geometría y Física del espacio-tiempo de Minkowski. La Gaceta **20**(3), 539–562 (2017)
97. S. Xambó-Descamps, *Real Spinorial Groups—A Short Mathematical Introduction*. SBMAC/Springerbrief (Springer, New York, 2018).
98. C. Yu, M. Jin, H. Liu, An analytical solution for inverse kinematic of 7-DOF redundant manipulators with offset-wrist, in *IEEE International Conference on Mechatronics and Automation*, Chengdu, 5–8 August 2012, pp. 92–97
99. J. Zamora-Esquivel, E. Bayro-Corrochano, Robot perception and handling actions using the conformal geometric algebra framework. Adv. Appl. Clifford Algebr. **20**(3), 959–990 (2010)
100. I. Zaplana, J. Lasenby, L. Basañez, General closed-form solutions for the inverse kinematics of serial robots using conformal geometric algebra. Mech. Mach. Theory, Under Revision (2018)

Index

Symbols
A^*, 42
BP, 103
C, 75
$\mathbb{C}$, 2
C, 11
$\mathcal{C}$, 33
CGA, v, 33
Conf(E), 48
$\mathcal{D}$, 23
  existence, 32
D_α, $\boldsymbol{D}_\alpha$, 46
DGP, 102
DMDGP, 103
$G^+(E)$, 47
$G(E)$, $\Gamma(E)$, 48
$\mathcal{G}_{1,3}$, 2
  existence, 32
  linear grading, 26
$\mathcal{G}_2$, 2
  existence, 32
$\mathcal{G}_3$, 2, 14, 15
  existence, 20
  linear grading, 16
GA, v, 1
GPS, 54
$\Gamma^+(E)$, 47
H, 36
$\mathcal{H}$, 36
$\mathbb{H}$, 19
H, 18
Id, 49
$J_A(\boldsymbol{q})$, 77
$J_G(\boldsymbol{q})$, 77
$k/\!\!/2$, 6
$\mathbb{N}$, 1
NMR, 101
$\bar{\mathrm{O}}^+$, 47
$\mathrm{O}(\bar{E})$, 44
O_2, 8
  structure, 13
$\mathrm{O}_2^\pm$, 12
O_2, 8
O_3, 14
$\mathrm{O}_3^\pm$, 14
O_n, 4
P, 13
$\mathrm{P}^\pm$, 12
$\mathbb{Q}$, 2
$\mathcal{Q}$, $\mathcal{Q}'$, 36
$\mathbb{R}$, 2
$R_{i\theta}$, $\boldsymbol{R}_{i\theta}$, 45
SO_2, 8
SO_3, 14
Spin_3, 19
T_v, $\boldsymbol{T}_v$, 45
${}^0T_n$, 76
V_s, 47
V_u, $\boldsymbol{V}_u$, $V_{u+\delta e_\infty}$, $\boldsymbol{V}_{u+\delta e_\infty}$, 47
$\boldsymbol{V}_s$, 48
X, 75
$Z = e_{0\infty}$, 35
$\mathbb{Z}$, 1
e_0, 34

C. Lavor et al., *A Geometric Algebra Invitation to Space-Time Physics, Robotics and Molecular Geometry*, SpringerBriefs in Mathematics,
https://doi.org/10.1007/978-3-319-90665-2

e_∞, 34
η, 23
$t(I, J)$, 24
$t(p)$, 25

A
algebra, 5
algorithm
  branch-and-prune (BP), 103
angle
  Euclidean, 4
  hyperbolic, 58
  oriented, 11
antiautomorphism, 6
Archimedes, 53
Artin's formula, 24
associative, 5
automorphism, 6

B
basis
  orthonormal, 4
bivector, 26
blade basis, 35
bond angle, 104
bond length, 104

C
Cartan-Dieudonné theorem, 14
Cartesian power, 5
Cauchy-Schwarz inequality
  Euclidean, 31, 58
  hyperbolic, 57
center, 17
charge
  conservation, 67
  density, 66
chronometry, 58
Clifford
  basis, 15, 25
  product, 9
  reduction rule, 9, 23, 33
  relations, 23, 34
  units, 15, 24, 34
complex
  scalars, 11, 30
  vectors, 30
$\mathrm{Conf}(E)$, 48
configuration space, 75, 76
conformal
  closure, 33
  compactification, 37
  geometric algebra, 33, 80, 83, 91
  reflection, 47
  vector, 36
continuity equation, 67
cross product, 21
curl, 64
  of vector, 66
current density, 66

D
D-H convention, 75, 80, 81
dalembertian, 64
decomposable k-vector, 5, 26
degrees of freedom, 77
Descartes, 53
differential, 64
differential kinematics, 83, 86
dilation, 46
Dirac, 53
  bispinor, 70
  current, 72
  equation, 53, 71
  field, 71
  gamma matrices, 71
  operator, 64
  representation, 23
  spinor, 70
directional derivative, 64
distance geometry, 102
divergence, 64
double cross product, 22
dual blade, 51
duality theorem, 42

E
Einstein, 53
  formulas, 70
  summation criterion, 56
electric field, 66
electrical permittivity, 54
electrodynamics, 53
electromagnetic field, 70
  energy, 73
  transformation of, 70
Elsewhere, 57
end-effector, 75
  pose, 76
Euclid, 53
Euclidean
  geometry, 53
  group, 48
    proper, 46

- metric, 4
- plane, 2
- reflection, 47
- space, 2, 4, 53
- vector space, 53

Euler, 53
- angles, 95
- formula, 31

even geometric algebra, 11
even subalgebra, 18, 29
event, 56
exterior
- algebra, 6, 98
- power, 5
- product, 11, 17, 26, 34

F

Faraday bivector, 66
flats, 51
forward kinematics, 77, 81
frame, 56

G

geometric
- covariance, 2, 20, 45
- product, 9
- quaternions, 19
- transformation, 44

grade, 6, 26
graded algebra, 6
Gram's formula, 26
graph
- simple, 101
- undirected, 101
- weighted, 101

Grassmann, 53
- algebra, 6

H

Hamilton, 19
Hestenes map, 36
Hestenes, D., vi, 71
Hestenes-Dirac equation, 71
Hodge duality, 17, 29
horosphere, 36
hyperbolic
- angle, 58
- cosine theorem, 58
- plane, 33
- triangle inequality, 58

I

identity matrix, 5
imaginary sphere, 39
inertial frame, 56
inner product, 21, 26, 34
inner representation, 38, 92
- of a circle, 40
- of a line, 41
- of a plane, 40
- of a sphere, 39

internal coordinates, 104
interval, 56
inverse kinematics, 77, 86, 88, 97
inversion, 48
involution
- parity, 6, 11, 17, 28, 34
- reverse, 6, 11, 17, 28, 34

IPNS, 38
isometry, 4
- group, 55
- orthochronous, 55
- proper, 55

J

Jacobian matrix, 77
- analytical, 77
- geometric, 77
- geometric Jacobian, 86, 97

joint, 75
- prismatic, 75
- revolute, 75
- variables, 75

K

k-blade, 5, 26
k-vector, 5, 26, 34
k-volume, 5
key formula, 11, 17, 27
kinematic function, 76
Klein's geometry, 55
Klein-Gordon equation, 71

L

lab representation, 65
Lasenby, A., vi
light cone, 57
linear map, 3
links, 75
Lorentz, 53
- bivector, 61

Lorentz (*cont.*)
boost, 54, 62
factor, 55, 65
force law, 67
invariant, 67
relativistic force law, 67
sphere, 57
transformations, 53, 54
Lorentzian vector space, 53
Lorenz gauge condition, 68

M
magnetic field, 66
magnetic permeability, 54
matrix algebra, 5
Maxwell, 53, 66
equations, 66
meet, 93
method
closed-form, 78
Jacobian-based, 78
damped least-squares, 79
pseudoinverse, 79
transpose, 79
numerical, 78
Paul's, 78
Pieper's, 86, 97
metric, 3, 53
alternative form, 28, 34
Michelson-Morley experiment, 54
Minkowski space, 2, 23, 53
mixed product formula, 22
motor, 80
multiindex, 24
multivector, 6, 16, 26, 34
multivector field, 64

N
Newton, 53
null cone, 36
null vector, 36, 91

O
operational space, 75
OPNS, 38
optics, 54
orientations, 11, 17
oriented
k-volume, 5
angle, 11
area, 5
volume, 5
orthogonal
basis, 4
group, 4
special group, 8
vectors, 4
orthonormal basis, 4, 23
outer (=exterior) product, 11, 17, 26, 34
outer representation, 41
of a circle, 42
of a line, 42
of a plain, 43
of a point pair, 41, 93
of a sphere, 43

P
parity involution, 6, 11, 17, 28
Pascal, 53
path
lightlike, 60
spacelike, 59
timelike, 58
uniform, 59
Pauli
algebra, 30
matrices, 21
representation, 20
spinors, 70
perpendicular, 4
photons, 60
Plücker coordinates, 98
planar robot, 81
Poincaré, 53
lemma, 68
polarization indentity, 3
pose, 75, 86
positive definite, 4
potential, 68
scalar, 69
vector, 69
Poynting vector, 73
principle of relativity, 54
projective
matrix, 81
space, 7
proper
distance, 59
time, 59
pseudoscalar, 26
pseudoscalars, 11, 17, 24
Pythagoras, 53

Q

quadratic form, 3
quaternions
  geometric, 19
  Hamilton, 19

R

rapidity, 55, 62
reciprocal frame, 63, 87
reflection, 14
  versor form, 18
relative
  representation, 65
  space, 29
  velocity, 65
relativistic
  chronometry, 58
  composition of velocities, 62
  electrodynamics, 53
  kinematics, 58
  mass, 66
  moment, 65
  rest mass, 65
reverse involution, 6, 11, 17, 28
Riemann, 53
Riesz, 66
  –Maxwell equation, 66
  method, 15
Rodrigues, 19
  formulas, 19, 22
rotation, 45
  axis, 18
  elements of, 18
  in a plane, 8
  rotor of, 18
  spinorial form, 18, 87
rotations, 14
  composition, 19
  in a plane, 12
rotor, 18, 60, 80, 88, 94
  condition, 18
  field, 72
rounds, 51

S

scalar separation, 56
scalars, 3, 9
Schrödinger, 70
  equation, 71
secondary task, 78
separation, 56
serial robot, 75
  non-redundant, 77
  redundant, 77
signature, 23
similarity transformation, 47
  proper, 47
singularity, 79, 98
  orientation, 80, 99
  position, 80, 99
  problem, 98
  representation, 77
skew-commutative, 6
special relativity, 55
speed of light, 54
spherical wrist, 78
spin
  bivector, 72
  vector, 72
$Spin_3$, 19
spinorial map, 13
spinors, 19
  3D, 19
  Dirac, 70
supercommutative, 6
symmetric difference, 24
symmetry, 8
  in a plane, 8, 12

T

temporal orientation, 57
torsion angle, 104
translation, 45
trivector, 26

U

unit vector, 56
unital, 5
universal
  constant, 54
  property, 6

V

vector, 9, 26, 56
  axial, 22
  field, 64
  future-oriented, 57
  isotropic, 23
  length, 4, 56
  lightlike, 56
  magnitude, 56
  negative, 23, 56
  norm, 4, 56

vector (*cont.*)
  normalization, 4
  null=isotropic, 23, 36, 56
  past-oriented, 57
  polar, 22
  positive, 23, 56
  Poynting, 73
  signature, 56
  spacelike, 56
  timelike, 56
vector separation, 56
vector space, 3
  basis, 3
  Lorentzian, 53
versor isometries, 47
volume element, 17, 24, 26

W
wave
  equation, 68
  function, 70

Printed in the United States
By Bookmasters